# War and the City

# War and the City

## Urban Geopolitics in Lebanon

Sara Fregonese

**I.B. TAURIS**
LONDON • NEW YORK • OXFORD • NEW DELHI • SYDNEY

I.B. TAURIS
Bloomsbury Publishing Plc
50 Bedford Square, London, WC1B 3DP, UK
1385 Broadway, New York, NY 10018, USA

BLOOMSBURY, I.B. TAURIS and the Diana logo are trademarks
of Bloomsbury Publishing Plc

First published in Great Britain 2020
Reprinted 2020

Copyright © Sara Fregonese 2020

Sara Fregonese has asserted her right under the Copyright, Designs
and Patents Act, 1988, to be identified as Author of this work.

For legal purposes the Acknowledgements on p. viii constitute
an extension of this copyright page.

Cover design: Adriana Brioso
Cover image: '20 Years' (oil on canvas) by Tom Young www.tomyoung.com

All rights reserved. No part of this publication may be reproduced or
transmitted in any form or by any means, electronic or mechanical,
including photocopying, recording, or any information storage or retrieval
system, without prior permission in writing from the publishers.

Bloomsbury Publishing Plc does not have any control over, or responsibility
for, any third-party websites referred to or in this book. All internet addresses
given in this book were correct at the time of going to press. The author and
publisher regret any inconvenience caused if addresses have changed or sites
have ceased to exist, but can accept no responsibility for any such changes.

A catalogue record for this book is available from the British Library.

A catalog record for this book is available from the Library of Congress

ISBN: 978-1-7807-6714-7
ePDF: 978-1-8386-0052-5
eBook: 978-1-8386-0053-2

Series: International Library of Human Geography

Typeset by Integra Software Services Pvt. Ltd.
Printed and bound in Great Britain

To find out more about our authors and books visit www.bloomsbury.com
and sign up for our newsletters

# Contents

| | |
|---|---|
| List of Illustrations | vii |
| Acknowledgements | viii |
| On translation and transliteration | x |
| List of militias mentioned in the book | xi |
| List of participants from the 1975–1976 war | xii |
| | |
| Introduction | 1 |
|    Scope of the book | 2 |
|    Anti-geopolitical eyes, subjugated knowledges | 4 |
|    Structure of the book | 7 |
|    Note on methodology | 9 |
| 1  Cities in the Space of Global Politics | 13 |
|    Contested histories | 14 |
|    New urban wars? | 19 |
|    Urbicide: Theorizing political violence against the urban built fabric | 22 |
|       A genealogy | 22 |
|       The city of urbicide | 24 |
| 2  Modernity, Territory and Conflict in Lebanon | 29 |
|    The sect as laboratory of modernity | 29 |
|    The refuge and the battleground | 33 |
|    The sect as territory. The Double Kaymakamate of Mount Lebanon (1842–1860) | 37 |
|    The sect as dispositif. The Mutasarrifiyya of Mount Lebanon (1860–1920) | 42 |
|    Normalizing the sect: The French mandate and the Lebanese Republic (1920–1946) | 47 |
|    Overlapping territories. The mountain and the shifting urban geopolitics of Beirut | 50 |
| 3  Lebanon Salvaged: Sovereignty and Urban Space in the Republic of Lebanon (1943–1975) | 55 |

|   |   |   |
|---|---|---|
|   | Geopolitical scripts of fragility | 57 |
|   | The urban geopolitics of Operation Blue Bat | 63 |
| 4 | Towards War | 67 |
|   | The missing link: Revisiting socio-economic perspectives on violence in Lebanon | 68 |
|   | Hybridizing sovereignty: The relationship between state and irregular armed groups | 71 |
|   | Urban architectures of enmity | 76 |
|   |     The Sidon fishing protests | 77 |
|   |     The bus shooting and the start of the war | 79 |
|   | The Two Years' War | 80 |
| 5 | Lebanon Lost: The Urban Impact of Non-Intervention | 85 |
|   | The tragedy script: Non-intervention as *territorial trap* | 88 |
|   | The chaos script: The politics of loss | 93 |
|   | The fate script: The unavoidable sect | 96 |
| 6 | De-Subjugating Beirut's Urban Geopolitical Knowledges | 101 |
|   | The built environment and the propagation of violence | 102 |
|   | Reframing sovereignty in the city | 108 |
|   | Geopolitical architectures: The battle of the hotels | 115 |
| 7 | Beirut's Hybrid Sovereignties: The May 2008 Clashes | 127 |
|   | From lack of sovereignty to hybrid sovereignties | 128 |
|   | Beyond the state/nonstate dialectic | 130 |
|   | Sovereignty and the built environment in the May 2008 clashes | 131 |
| Conclusion |   | 137 |
|   | De-subjugating geopolitics | 138 |
|   | Decolonizing sectarianism | 140 |
|   | Contextualizing urbicide | 141 |
|   | Towards pacific urban geopolitics | 142 |
|   | Beirut's possible spaces for peace | 144 |
| Notes |   | 148 |
| References |   | 157 |
| Afterword |   | 175 |
| Index |   | 180 |

# List of Illustrations

1. Map of the present seat of war in Turkey and Syria. 1840. British Library, Maps 46970.(6.)    38
2. The Lebanon. Divided into Kaimakamiyes [sic] according to the Commissioner's 1st project of XLVII Articles. 1861. The National Archives, Kew. FO 925/2889    41
3. Rasheya [Rashaya, Lebanon], 27 April 1862. By Francis Bedford (1815–94). Acquired by the Prince of Wales (later King Edward VII), 1862. Obtained with permission by the Royal Collection Trust. Catalogue No: RCIN 2700955    45
4. Political poster by the Arab Socialist Union. American University of Beirut/Library Archives. Political Posters. No. 177-PCD2081-25    61
5. US Marines intervention in Lebanon. Cartoon by Michael Cummings, The *Daily Express*, 9 July 1958    65
6. Lebanese Forces propaganda poster. American University of Beirut/Library Archives. Political Posters, No. 354-PCD2709-06    112
7. Al-Murabitun propaganda poster. American University of Beirut/Library Archives. Political Posters, No. 158-PCD2081-17    120
8. Al-Murabitun propaganda poster. American University of Beirut/Library Archives. Political Posters, No. 159-PCD2081-16    122

# Acknowledgements

This book results from countless interactions with people, institutions and places. I would like to thank the editors at IB Tauris and Bloomsbury Publishing for their professionality and patience. Funding allowing the research for this book stems from a PhD studentship at the School of Geography, Politics and Sociology at Newcastle University (2004–2007), a British Academy Post-Doctoral Fellowship (PDF/2009/428, 2009–2012) on state-nonstate interactions in urban conflict, and a British Academy Small Research Grant (SG102042, 2011–2012) into the historical geographies of cosmopolitanism through Beirut's former hotels. The research in Chapter 2 stems from a more recent CBRL Pilot Study Award (2016) into the historical geography of sectarianism in late-Ottoman Lebanon. The research was presented at workshops, seminars and conferences in Europe, United States and the Middle East, including Cambridge University, Royal Holloway University of London, Royal Geographical Society, University of Manchester, Ghent University, University Of Oxford, University Of Zurich, International Boundaries Research Unit (IBRU) School of Advanced Study (London), American Association of Geographers, Orient Institut Beirut, Lebanese American University, and The Netherlands Institute in Turkey.

To the research participants, especially those in Chapter 6, who conceded me lengthy interviews to talk about their daily lives and experiences of war and the city, goes my deepest thank you. I am indebted to the residents, workers, journalists, party and organization representatives, architects, urbanists, and academics who engaged in interviews and in countless conversations with a purpose during my stays in Lebanon.

The hospitality of several Beirut-based institutions was essential to conducting the research. I am thankful to Orient Institut Beirut (OIB), the Centre for Arab and Middle Eastern Studies (CAMES) and the Prince Alwaleed Bin Talal Bin Abdulaziz Alsaud Center for American Studies and Research (CASAR) at the American University of Beirut (AUB). During fieldwork, I benefitted from the use of libraries and archives including: Institut Français du Proche-Orient (IFPO), AUB's Jafet Library's Archives & Special Collections, the Library of the Campus de Sciences Humaines of the Universite' Saint Joseph (USJ) and USJ's Librairie Orientale, Al Nahar Documentation Centre, Umam Research and Documentation Centre, The National Archives in Kew (London), the Prime Minister Archives in Istanbul, the online archives of the Ford Library and the online French diplomatic archives.

## Acknowledgements

I am deeply grateful to those who assisted me with fieldwork research, translation and transcription: Waleed Serhan, Simona Loi, Imad Aoun and Ayhan Han. I am indebted to Beirut-based artist Tom Young for his permission to use his artwork for the front cover.

I am grateful to Claudio Minca, Alex Jeffrey and Alison Stenning, my doctoral supervisory team at Newcastle University, who provided outstanding scholarship, indefatigable draft-reading and friendship. The School of Geography, Politics and Sociology has been a welcoming environment and in its postgraduate community I found some of my best friends. Klaus Dodds has been an academic and life mentor at RHUL between 2009 and 2012, and commented on a substantial number of previous drafts on which this book builds.

Many friends have been inspirational through personal, scholarly and fieldwork-related conversations, reading drafts, offering support and hospitality, and accompanying me during long research walks across Beirut. My deepest thanks go to Dana Abi Ghanem, Luiza Bialasiewicz, Federico Calabrese, Estella Carpi, Maria Kastrinou, Caroline Nagel, Wendy Pullan, Nadim Shehadi, James Sidaway, Lynn Staeheli, Lorenzo Trombetta and to the many more who have helped at various stages of the research underpinning this book.

My utmost gratitude goes to my family. To Adam, for unwavering presence, infinite patience, lifesaving IT savviness, piano background music provision and for being an outstanding parent for Layla throughout my work on the manuscript. And to our daughter Layla, who has *not* been patient while I wrote this book, and quite rightly. She has taught me how a book can be written while working part-time, without sacrificing many hours of family life, and without confusing overwork with commitment.

Sara Fregonese
July 2019

# On translation and transliteration

Technical and context-specific words and historical place names in non-Latin alphabets have been transliterated into English, following the translation and transliteration guidelines provided by the *International Journal of Middle Eastern Studies* (IJMES).

Current names of places follow the accepted English spelling (e.g. Beirut, Sidon). Names of prominent personalities and parties also follow the accepted spelling in the English press, with which readers might be more familiar.

# List of militias mentioned in the book

## Lebanese National Movement (Al-Harakat al-Wataniyya al-Lubnaniyya)

Independent Nasserite Movement (Al-Mourabitun)
Arab Socialist Union
Lebanese Communist Party
Popular Nasserist Organisation

## Non-Lebanese allies of the National Movement

Democratic Front for the Liberation of Palestine (DFLP)
Popular Front for the Liberation of Palestine – General Command (PFLP-GC)
As-Saiqa

## Lebanese Front (al-Jabhat al-Lubnaniyya)

National Liberal Party (Al Ahrar)
Lebanese Phalanges (Al Kata'ib)
The Organization (Al-Tanzim)
Marada Current
Guardians of the Cedars

# List of participants from the 1975–1976 war

George Corm wrote that, during the civil war in Lebanon, 'the last of the militiamen [...] could become a historian and enact history' (Corm 2005, 41) as the result of a tendency among both the political elites and lay citizens to envision Lebanese history outside the machinations of the great global or regional powers. The militia combatants, according to Corm, very often contextualized – and legitimized – their actions within a wider geopolitical narrative, in line with this or that superpower's agenda.

Despite a general amnesty for crimes perpetrated in Lebanon between 13 April 1975 and 28 March 1991, caution has persisted in making public the experiences of militia fighters, in a country where there is no state-driven truth and reconciliation initiative and where an agreed version of civil war history remains highly contested (Launchbury 2014). Barring few exceptions (Duplan and Raulin 2015; Saadé, et al. 1989), it is only recently that former fighters have started a more sustained dialogue around the war, with grassroots initiatives like *Fighters for Peace*.

As an urban account of war, this book relies, among other materials, on firsthand experiences of combat as recalled by five former combatants of the first phases of the war, who agreed to share their testimonies with me and my research assistants. Chapter 6 is based on their personal experiences of fighting on the urban ground, in the name of wider territorial and political projects. In order to contextualize their statements in Chapter 6, I provide here short segments of their lives, derived from salient details in the interviews. For the reasons above, I promised anonymity even to those who didn't mind their name being used. Below, I replaced their names with fictitious ones, inspired by Lebanese playwright Ziad Rahbani's theatre piece *Film Ameriki Tawil*.[1]

## Abu Layla (31 October 2005)

Abu Layla is a soft-spoken man. He was fourteen when the war started and spent the entirety of the conflict in a neighbourhood situated by the demarcation line. Despite being apolitical, he had friends and acquaintances among the fighters. He initially guarded one of the demarcation points by the Kataeb headquarters near the city centre. After participating directly in combat, he then worked in the media department of the Lebanese Forces (al-Quwwāt al-Lubnānīya). Abu Layla considers himself a good shooter because he used to shoot birds in the mountains before the war. He would like to write a memoir. What saddens him most is thinking of his schoolmates: ten out of twenty died in the war, the exact half.

**Omar (2 November 2005)**

Omar fought during the entirety of the civil war. He was a National Movement field commander and in 1975–1976 he supervised several positions along the confrontation line, from the centre of Beirut to the outskirts and as a consequence he moved between positions across the city. He was responsible for planning attacks and coordinating operations over the radio. Before the war, he used to go the cinema in the St Charles centre where the Holiday Inn was located, although he viewed this building as a symbol of bourgeois capitalism. As a hobby he paints scenes of local elections.

**Edouard (22 November 2005)**

Edouard trained with the Lebanese Phalanges party (*Ḥizb al-Katā'ib al-Lubnānīya*) in the early 1970s when he was a student, hiding it from his parents. On 13 April 1975, he was in the mountains with his parents. Upon returning to Beirut, he reached by car the frontlines of the first rounds of fighting. He fought the battle of the hotels and spent twenty-four hours in the Holiday Inn before it was taken over by the militias belonging to the National Movement. He left Beirut to go abroad in 1979 and returned to combat in 1983. He says that his war was against the Palestinians and the Syrians, but the day he realized that the Lebanese were fighting among themselves, he understood that the war was lost. Edouard bangs on the table when he speaks about the Palestinians and the Syrians.

**Nizar (1 December 2005)**

Nizar lived in the Al Shiyyah neighbourhood in south Beirut and was sixteen at the beginning of the war. He used to work during the day and study in the evening. On Sunday 13 April 1975, a warm day, he was swimming in the nearby seaside resort of Jounieh with a friend. On the bus back to Beirut, as they entered their neighbourhood, they found out that another bus had been shot at in the adjacent area of Ayn al-Rummana. The bus driver sent everyone off and they went home. He didn't know much about politics, but when he started training with the National Movement at the beginning of the war, he started speaking politics. As a resident of Al Shiyyah, his combat position was in the area. He was also involved in the hotels battle. Nizar doesn't like violence. He thinks the Lebanese are all children of the same country.

**Rashid (1 December 2005)**

Rashid was twenty-one when the war started. He lived in a neighbourhood right on the demarcation line. He says that even those who didn't know about politics knew that Lebanon was going to war, because militants from all sides were training with weapons. In the Two Years' War, Rashid fought with the National Movement, because his house was on the confrontation line. He was completely dedicated and quit his new business in order to be a fighter. In the militia, he was responsible for the area of Khand el Ghamiq, but participated also in confrontations in nearby areas. He spent six months on the frontlines of the battle of the hotels, during which he was shot twice.

# Introduction

War in Lebanon, from 1975 to 1990,[1] involved heavily the material built fabric of the capital Beirut (Davie 1983; Labaki and Abou Rjeily 1993). During my first trips to Lebanon as an undergraduate student of Arabic language and writing a dissertation on the influence of postmodernism on Beirut's postwar reconstruction, I used to listen to wartime stories with fascination for at least three reasons. Firstly, these stories involved very minuscule portions of urban space: roads, squares or single buildings or even specific portions of buildings. Secondly, these spaces, despite their small dimensions, acquired very substantial political and religious meanings. Thirdly, these tales differed starkly from the macro-geopolitical accounts of the press and official diplomacy, more often than not depicting Lebanon's civil war as a proxy battleground for regional geopolitics. This book results from that initial fascination: it moves those 'non official' and urban views and experiences of the war to the frontstage of enquiry. The narrations in this book reveal a differential and micro-level view of the civil war and of how the city was transformed by it. The aim is to shift away from the macro-analyses and instead propose a micro-level exploration of urban conflict that repopulates those formal geopolitics with actual bodies, spaces, and materialities.

In his book *Beirut* the late historian Samir Kassir (2010) poignantly noted how Beirut is not an 'original city' (440), a passive victim of conflict coming from nowhere. He argues that much of the political and (para) military mobilization towards the civil war was obtained by 'transposing the convulsive upheavals of regional geopolitics to the urban level' (Kassir 2010, 440). This book focuses on that transposing of regional issues onto the urban ground. It traces the connections that link large-scale geopolitical discourses about Lebanon with the micro-geographies of violence in and against urban sites during a defining urban moment of the conflict: the war of 1975–1976, known widely as the Two Years' War (*al-harb as-sanatayn*). The following chapters offer an in-depth account of the relations between geopolitics, armed conflict and Beirut's urban space, and of how Lebanon's sovereignty has been implicated in and reshaped by these relations. The book also traces the deeper historical and territorial connections between the urban impact of the Two Years' War and Lebanon's violent colonial past, questioning the very nature of its current sovereignty in light of contested histories of violence.

The questions around which the book revolves are: firstly, how do large scale, formal geopolitics and the microgeographies of conflict interact and intersect in the representation and the practice of war in Lebanon? Secondly, how was the urban built environment of Beirut involved in the processes of production and mobilization of those representations and practices? Thirdly, what are the legacies of the histories and territories of Lebanon's Ottoman and colonial past on its urban geopolitics of conflict? Fourthly, what does an urban perspective on conflict tell us about the nature of sovereignty in Lebanon?

## Scope of the book

Several studies of the consequences of the civil war on the urban geography of Beirut consider the entire period of the war from 1975 to 1990 (Al-Harithy 2010; Calame and Charlesworth 2009; Khalaf and Khoury 1993; Nagel 2002; Sawalha 2010), even though its spatial impact changed remarkably throughout different phases, as a result of shifting political, military and paramilitary alliances, of the intervention of foreign state armies and the introduction in the conflict of different military technologies like air power. Lebanon scholars have also produced several sophisticated understandings of the Lebanese conflict, and especially of the nonstate armed actors that enacted de facto sovereignty in the Lebanese capital during the war. They have investigated their birth and the development of their mechanisms of violence (Corm 2005; F. El-Khazen 2004; Hanf 2015), their role as providers of social services replacing the state (Harik 1994), and as political actors crossing sectarian divisions in order to realize wider political projects (Rowayheb 2006). Other authors have focused on the relation between militia territoriality and visual mechanisms of propaganda (Chakhtoura 2005; Maasri 2008), and the power structure and organization of specific militias and the production of specific spaces (Kemp 1983). The value of these studies resides in their breadth, as some consider the whole duration of the war; or in their focus on the features of one single militia or on one specific aspect of the militia activity (such as public services). However, this previous work often does not, or only partially, consider the built environment as part and parcel of the production of discourses of war, sovereignty and identity.

The scope to this book is less wide, but explores more deeply the relationship between war and urban built environment during the crucial early phases of the civil war. As I started my PhD, I intended to examine how regional geopolitics intersected with and manifested themselves in the urban politics of a post-conflict urban environment like that of Beirut. Instead, as the research moved on and as a lively debate emerged in urban geopolitics in 2004 and 2005, I chose to focus my attention on the initial phase of the conflict between 13 April 1975 and 21 October 1976. I apply instead interpretive depth to understand how geopolitical meaning is renegotiated *during* conflict, and especially during close-quartet urban warfare with light or relatively light weapons, where urban streets and buildings compose both the fabric that shapes the rationalities of war, and the materiality through which meanings about sovereignty, territory and nation state were renegotiated and re-inscribed. The Two Years' War became my field of enquiry into the urban geopolitics of wartime Lebanon.

Most specialized literature on urban space in Lebanon deals with post-war reconstruction, leaving the crucial phase of *destruction* under-analysed. One of the reasons for choosing to concentrate on the Two Years' War concerns the different ways in which space was produced through violence in the pre- and post-1978 phases of the war. Various scholars (Corm 2005; Kassir 2003) agree about a change in the strategic politics, geographical span, and geopolitical weight of the Lebanese conflict after March 1978, when the Israeli army launched Operation Litani.[2] They propose that 'beginning in mid-1978, the nature and scope of the war changed' (Farid El-Khazen 2000, 5): the entry of the Israeli armies on Lebanese soil (and, in 1976 of the Arab Deterrence Force led by Syria); the use of heavier weapons, air power and siege (in 1982); the creation of the international peacekeeping force; and finally the rise of suicide terrorism in the 1980s have very often been considered as dynamics that shifted the Lebanese conflict onto a different scale. This shift included the internationalization of the conflict, the solidification of the militias' structure and organization (El-Khazen 2000) and of their affiliations to foreign powers (Corm 2005), as well as the increasing use of heavy artillery and the use of air power.[3]

The relations between aerial warfare and the spaces that they produce have been widely investigated (Gregory 2006; Hewitt 1983; Lindqvist 2002). There are profound material and epistemological differences between asymmetric urban guerrilla warfare versus airpower. In the latter, for example, the reliance on the objectivity of military technologies turns the urban terrain into a detached and sanitized blank space, reducing the bodies of the victims to 'disembodied abstractions' (Gregory 2004, 54) like points on a map. The use of air power by the IDF in Lebanon also bears these material and cultural implications, the analysis of which – although much needed – falls beyond the scope of this book.

Aerial power impacted materially onto Beirut's urban fabric differently from the material imprint of the previous phase, mainly because of the vast destruction of extended air attacks and heavy artillery, which substituted or were superimposed on the irregular and partial ruination of close-quarter and relatively lightly armed guerrilla fighting (Möystad 1998) of the previous phase. Most importantly, the rounds (*Jawlat*) of fighting that punctuated the Two Years' War were crucial in establishing the frontlines and, ultimately, the partition of the city into two sectors (East and West) until the end of the conflict.

## Anti-geopolitical eyes, subjugated knowledges

Critical and post-structuralist theory has dismantled the claims of objectivity and scientific nature of geopolitics and has highlighted instead the contested mechanisms through which understandings of security and global politics are produced contextually (Campbell 1998). Dijkink describes Geopolitics as 'particular but shared visions (narratives) of the meaning of one's place in the world and the global system' (1996, 1). These narratives, according to Dijkink, are discourses working to link the history and values of a political entity (the nation state) to a specific territory, legitimating specific visions through repetition, reference to tradition and the employment of recognizable tropes in order to build narratives about the rest of the world and turning them into common sense.

On the same line, *critical geopolitics* has analysed the ways in which knowledge about international statecraft is produced, represented and translated into practices of foreign policy (Ó Tuathail 1996; Dodds and Sidaway 1994; Tuathail and Agnew 1992). It has unpacked how specific geopolitical visions are produced contextually, rather than assuming that they are objective depictions of the world. It is argued by critical geopolitics scholars that 'government ministers or political figures are not outside the hegemonic national culture and thus the prevailing circulation of ideas and values shape their pronouncements' (Dodds 2003, 128). These metaphors then translate into mappings (the Axis of Evil, the Clash of Civilisations or the Failed State Index, for example) that inform the practice of global foreign politics, providing the rhetorical foundations to support state policies and (military) action. In the words of Derek Gregory, these mappings shape global 'architectures of enmity' (Gregory 2004).

This book utilizes two specific notions from critical geopolitics scholarship to address Lebanon's civil war and its relation to urban space, and the genealogies of sovereignty imbricated in this relation. The first idea is the notion of 'anti-geopolitical eye' (Ó Tuathail 1996a). It consists in 'a disturbing way of seeing that disrupts the framework of the hegemonic geopolitical eye that structures the seeing of places ... in contemporary foreign policy discourse' (1996a, 173). The embodiment of the anti-geopolitical eye, according to Ó Tuathail, was British journalist Maggie O'Kane's reporting from the war in Bosnia in the early 1990s. Through her proximity to the conflict scene and her emphasis on the voices and bodies of the civilian victims of the war, rather than on the official foreign policy statements, and personalities, her journalism posed a challenge to the detached view of the 'experts' analysing and commenting on international politics news.

According to Ó Tuathail, the anti-geopolitical eye should not negate geopolitics or aim to constitute a truer version of it. The anti-geopolitical eye should instead unsettle dominant geopolitical narratives by exposing differential narratives that had remained 'morally invisible' (1996a, 220). The idea of an anti-geopolitical eye, in other words, should not lead to a competitive dialectic between dominant and differential accounts. We should not claim truth out of geopolitical counter-perspectives, but rather expose other, alternative geopolitical imaginations alongside dominant ones, and the power/knowledge mechanism that structured their relationship.

The second idea is Michel Foucault's concept of 'subjugated knowledges' (Foucault 2003, 7), on which the notion of anti-geopolitical eye partly builds. Michel Foucault defined these as 'the non commonsensical knowledges that people have' (2003, 8). Bearing in mind the inextricability of power and knowledge,[4] in Foucault's thought, a knowledge is subjugated when it has not acquired or produced what constitutes legitimate recognition in a specific episteme, so to become 'common' sense. It is, therefore, 'a particular knowledge, a knowledge that is local, regional, or differential, incapable of unanimity and which derives its power solely from the fact that it is different from all the knowledges that surround it' (2003, 7–8). The notion of subjugated knowledges stems from Foucault's (Foucault 2000, 354) view of power – and especially state power – not so much as an utterly dominating force imposed onto society from an univocal external authority (what Foucault names negative power), but rather as a complex and creative web of heterogeneous manifestations. These manifestations spring from the dialectic (positive power) between the attempt to impose a will and the tactics to resist, reinterpret and renegotiate it. Rather than irradiating from a unitary source, for Foucault, power is better illustrated as a constellation of technologies acting pervasively in different localized contexts, continuously changing form and specifications:

> There exists no single power, but several powers. Powers, which means to say forms of domination, forms of subjection, which function locally [...]. All these are local, regional forms of power, which have their own way of functioning, their own procedure and technique. All these forms of power are heterogeneous. We cannot therefore speak of power, if we want to do an analysis of power, but we must speak of powers and try to localize them in their historical and geographical specificity. (Foucault 2007, 156)

In Foucault's vision, the 'techniques' (Foucault 2000, 357) through which power works are immanent to the social context in which they operate and evolve in, acquiring 'specific effects' (2000, 362) at the local level and through the practices of localized subjects.

Foucault's (2007) 'positive' notion of power as localized, dynamic and locally specified is crucial for opening the ground to the differential (or subjugated) geopolitical knowledges of Lebanon's Two Years' War, and especially those that developed from the urban battleground and that de facto renegotiated the predominant geopolitical visions of Lebanon, its territory and identity. The point of exposing these knowledges is not to make them the only true (or 'truer') ones, but rather to expose and subvert the mechanisms of power that allow one or the other to become dominant.

The notion of subjugated knowledges resonates with Edward Said's idea of contrapuntal reading of the literary source (Said 1994). A contrapuntal reading of seemingly distant sources analyses the zones of epistemological contamination between both. Edward Said (1994) borrowed the expression from the musical vocabulary, where counterpoint indicates a melodic combination obtained by accompanying a given melody with a different but complementary one. Said turns the idea of counterpoint into a theory and a methodology to critically de-subjugate the absent and obfuscated aspects that are nevertheless implied in the discourses and practices of imperialism.

In sum, this book offers an anti-geopolitical eye to normative geopolitical visions of the early stages of the civil war in Lebanon and in the capital Beirut. It is the result of listening closely and at length to the oral histories of militia combatants, and of archive research into the chronicles of the battles and the militia representations of the war, to highlight the geopolitical visions that were produced in a raging city during a more or less systematic, and extremely violent, process of destruction and spatial partition. The anti-geopolitical eye allows to form a counterpoint between the hegemonic and the subjugated geopolitical knowledges of the Two Years' War, and thus unpack the actual process of emergence, consolidation and material specification of geopolitical visions about Lebanon in 1975 and 1976.

## Structure of the book

The last decade has seen an increase in scholarly literature on the importance and role of cities within dynamics of violence, war and global politics. I review this literature and contextualize it within the political violence in Lebanon in Chapter 1, with particular attention to the interdisciplinary work in *urban geopolitics* (Bialasiewicz 2015; Fregonese 2012b; Graham 2004b) and the idea of *urbicide* as deliberate violent action against cities (Abujidi 2014; Berman 1984; Coward 2009; Huxtable 1989; Ramadan 2009; Tyner et al. 2014; Watson 2013).

Chapter 2 casts Lebanon's civil war within its wider historical context, by offering a territorial history of the emergence of Lebanon as a nation state where the religious community (Taifa) acted as governmental and territorial dispositif (Foucault and Gordon 1980) that became a foundation of the modern nation state. Sectarianism has been politically and territorially normalized between the 1840s after the Turco-Egyptian war and the 1920 institution of Greater Lebanon under French Mandate. The chapter goes beyond the scope of a historical review, and instead provides a reflection over the political, spatial and cartographic legacy of Lebanon's colonial politics. These are seen as key to understand Lebanon's more recent geographies of violence, connecting geopolitical changes in Ottoman and colonial Lebanon to the ideas around sovereignty, national identity and religion that circulated in Beirut during the Two Years' War and that were manifested through its built environment.

Views of what Lebanon's sovereignty entails and of what foreign policies a threat to its sovereignty justifies have changed in different phases of the country's modern history. Chapters 3 and 5 especially, focus on these understandings and on what has been included, changed, and silenced from those discourses around sovereignty. This allows us not only to pinpoint continuities and discrepancies in the understanding and practice of international foreign policy towards Lebanon. It also allows us to understand the mechanisms underpinning decisions about foreign intervention (or absence thereof) at different moments of intra-state conflict. Chapter 3 analyses the scripting of Lebanon as a place worth of military intervention through the US' Operation Blue Bat in 1958. In a tense Mediterranean region at the height of the Cold War, Lebanon's sovereignty was seen as something needing 'salvaging', in the name of the Eisenhower Doctrine, with a direct intervention of the US Marines into the inner neighbourhoods of Beirut. The chapter highlights the complex connections between the Cold War global geopolitical script and the localized urban and infrastructural aspects that shaped post-independence Lebanon.

Chapter 4 offers a spatial and urban viewpoint of the escalation towards the violence of the Two Years' War. This chapter questions how a country with generally positive socio-economic indicators and a rate of unequal development that was by no means unique and in any case not sufficient to explain the intensity of the violence descended into civil war in a matter of months. To do so, the chapter focuses on the spatial and material processes that manifested deep-seated controversies over sovereignty and security in Beirut and other towns. Here the use, control and reorganization of urban territory were the tools through which social and (geo)political difference was reframed as antagonism, and specific urban spaces were reinscribed as strategic, contentious or targetable.

Chapter 5 looks at how official international diplomacy approached the Two Years' War, and how these approaches scripted Lebanon's sovereignty in a way that, differently from 1958, discouraged intervention. The chapter argues that a macro-scale reasoning focused on the Cold War balance of power and détente geopolitical agendas failed to grasp or give importance to the complex urban context of increasing division and geographies of wanton violence inside Beirut, de facto allowing urban partition to happen. This complexity and gravity of the situation on the ground were repeatedly picked up by diplomats and even intelligence agents on the ground, but the official international government spheres overlooked those textured accounts of the conflict.

The complexity of urban fighting that was overlooked by the official international foreign policy discourses analysed in Chapter 5 is instead the focus of Chapter 6. Here, I account for the role of the built environment in the urban warfare in 1975 and 1976, tracing the strategic and political framings by armed militias and the impact of these framings on the incumbent partition of the city. Expanding empirically on the idea of urbicide, the chapter discusses the tactics of fighting on and through the built environment and how non-state actors produced differential geopolitical meaning (about sovereignty, and about the presence and role of the Lebanese nation and its role in relation to other nation states). These not only tell us a great deal about how armed non-state actors (re)made sense of the space of Beirut and reinterpreted its political geographies, but also show how these reinterpretations exerted remarkable power in shaping the process of urban partition.

The concluding chapter returns more specifically on the question of sovereignty in contemporary Lebanon. Drawing on the idea of hybrid sovereignties (Fregonese 2012a, b), the chapter takes the case of the deadly clashes of May 2008 in Beirut to show how both state and non-state actors continue to negotiate power over urban territory, built environment and infrastructure. In so doing, it re-frames Lebanon not as a 'weak' state where state sovereignty is lacking in light of the civil war, but as one where sovereignty has become increasingly *hybrid*.

## Note on methodology

This book is, at least partly, a historical account of the geographies of the war. Rather than a definitive chronicle of events, producing historical research is a negotiation between the nature and availability of sources and audiences, the positionality of the researcher (White 1987, 5), as well as the often messy and non-linear 'operation of memory' (Ballinger 2003, 5). History is not the straightforward result of unveiling historical pre-existing 'facts', and the production of history cannot be separated from the social process in which it is produced. Michel Foucault (2002) despoiled history of its aura of abstraction and immersed it into the worldly dynamics between power, knowledge, social processes and the documents that they constantly produce, approve or de-legitimize, so that 'the document is not the fortunate tool of a history that is primarily and fundamentally memory; history is one way in which a society recognizes and develops a mass of documentation with which it is inextricably linked' (Foucault 2002, 7). Whether we are dealing with nineteenth-century diplomatic correspondences about violence in Mount Lebanon, or with the fighters' narrations of the civil war, one is bound to deal with the methodological and political pitfalls of studying history in Lebanon. In 1991 a general amnesty decree dissolved the Lebanese of all liabilities for civil war crimes committed before March 1991. Amnesty and the absence of a national initiative of truth and reconciliation have led to the development of a kaleidoscopic and contested zone where myriad of different 'memory cultures' (Haugbølle 2010) exist, and where every religious and political group has its own version about their history and role in the war. Initiatives related to remembrance and responsibility are mainly in the hands of NGOs and private archives like *Fighters for Peace, MARCH Lebanon* and *Umam*. These initiatives are crucial not only in contributing to healing or closure (in whichever form) for individuals, but also in repopulating the memory of wartime Beirut with real bodies and concrete spaces, and providing indispensable counternarratives to the discourses of official geopolitics about the war. The empirical chapters engage contrapuntally with both official foreign policy documents and statements and with unofficial (and oral) militia accounts. This is not an attempt to dismiss official geopolitical narratives about Lebanon, or to declare the militia narratives as truer, but rather to trace their epistemological connections. And ultimately, to ground the reinterpretation and re-negotiation by non-state actors of official discourses of sovereignty and national identity onto the urban battleground.

The historical literature on late-Ottoman Lebanon draws on archive research conducted in 2016, 2017 and 2018 in London, Beirut, Baaqleen, and Istanbul in English, Arabic, French and Ottoman Turkish. It includes diplomatic correspondences, official treaties, petitions, demographic statistics and historical maps on the period between the end of the Turco-Egyptian war in 1839, and the declaration of the Mutasarrifiyya (Governatorate) of Mount Lebanon in 1860. The empirical material related to the civil war comes from official statements, parliamentary addresses, official government meeting minutes, and televised interviews related to foreign policy towards Lebanon of the United States, United Kingdom, France, Israel and the Vatican between April 1975 and November 1976. All documents are unclassified except for a number of passages in the White House Cabinet meeting minutes under US President Gerald Ford. Each of these statements is unique in its own way, and each of the states analysed had a particular foreign policy agenda (as well as a different historical connection) towards Lebanon.[5] The militia accounts were collected during fieldwork in Beirut between October 2005 and January 2006, and draw on biographical accounts of former militia fighters, all involved in different ways in the early rounds of the fighting in 1975 and in the battle of the hotels, which was crucial for completing partition of the city in 1976. These in-depth accounts explore the interconnections between war and urban space in what is still an understudied phase of the conflict. What were initially designed as semi-structured interviews soon turned into a form of storytelling. As conversation started to flow after the first questions, I often let my participant tell me their story: how they saw the war starting, how they became involved in 'political talk', how they trained, fought, how the militia was organized and most importantly how they related to the city they fought in (as a physical battleground but also as subjective experience). I very often had to ask for clarifications. Most importantly, on occasions their accounts referred heavily to the regional geopolitics, so I often found myself to metaphorically 'scale down' my interlocutor from these regional conjunctures back to the urban ground. I soon stopped letting this bother me and my 'research design', because the two are intertwined and there was no telling of one without the other. The most mundane of militia practices – from buying ammunition to passing by home to change clothes, to picking up the packed sandwiches from the local party headquarters – were all part of the same mechanisms of violence linked to specific spatial and political agendas. A mechanism that realized itself by acting against and through the urban space and built environment of Beirut. Further civil war material stems from archive research (text and microfilm) into the local press in French, English and Arabic, militia visual propaganda posters, and on countless field interactions that did not class as fully fledged interviews, but rather as conversations with a purpose with architects, journalists, academics, politicians and lay residents. Interviews with former hotel staff and archive research into newspapers, magazines, tourist guides, and advertisement material conducted during two short visits to Beirut in 2011 and 2012 provided the material on the battle of the hotels. Architects and planners have helped understand some of the processes informing architecture and society in mid-1970s Beirut. In wartime, local journalists' work to report on the urban battles often involved negotiation of access and coverage directly with militia members, thus providing further insight in the practices of fighting. Finally, visual material including graffiti, press photographs and historical imagery from political parties' websites, videoed interviews and statements by militia leaders provided additional context to the analysis. Finally, research into the May 2008 intra-state urban warfare between opposing armed groups (also known as the 7 May clashes) was conducted during a post-doctoral fieldwork visit from October to December 2010, consisting of twenty interviews with first-hand witnesses of the clashes, some of whom were inside targeted buildings during the clashes.

Parts of Chapters 2, 4, 5, 6 and 7 are based on reflections on the postcolonial nature of Lebanon's contemporary violence; the impact of the presence and role of Palestinian refugee camps on ideas and practices of sovereignty; the contrapuntal diplomatic and non-state knowledges and understandings of the civil war; and the notion of hybrid sovereignty that appeared, respectively, in articles published in *Geographical Review* (Fregonese 2012c) *Annals of the American Association of Geographers (*Ramadan and Fregonese 2017), *Political Geography* (Fregonese 2009), *Environment and Planning D: Society and Space* (Fregonese 2012a) and Geography Compass (Fregonese 2012b).

# 1

# Cities in the Space of Global Politics

The role of urban space in Lebanon's civil war can be conceptualized within wider theories of the contemporary geopolitical role of cities. According to these theories, with the end of the Cold War, interstate wars in open terrain have become less common and 'the informal, "asymmetric" or "new" wars that centre on localized struggles over strategic urban sites have become the norm' (Campbell et al 2007). While interstate conflicts have arguably disappeared, it is however becoming increasingly accepted that cities are crucial terrains of geopolitical contestation: 'From guerrilla warfare and terrorist attacks aimed at key installations, to coercive state occupation of strategic spatial locations … the terrain of everyday political struggle over who has the legitimacy to govern a people and on the basis of what identity categories seems to be scaling down from the nation-state' (Davis and De Duren 2011, 2) into the densely built fabric of cities, thus blurring the 'traditional binaries of war and peace, the local and the global, the civil sphere and the military sphere, the inside and outside of nation states' (Graham 2010, 294).

The War on Terror has been underpinned by materialities and technologies geared for complex and close-quarter urban combat situations, and by a takeover by cities of what has been traditionally considered as the geopolitical remit of the state: 'the defence of boundaries, the securing of territory, the management of flows of people and goods' (Bialasiewicz 2015, 317) as state practices of border control, securitization and mobility management are being implemented not only at state borders, but also within the physical infrastructure and built fabric of cities (Graham 2010; İşleyen 2018; Iossifova 2013). Yet, scholarly and practical accounts of conflict and sovereignty are still overwhelmingly preoccupied with the determination and management of territorial states and national borders. Far less academic work is dedicated to the exercise of conflict and sovereignty in and through urban environments *within* state borders, and for a long time there has been an 'almost complete dominance of national, rather than subnational [and urban], spaces and politics within International Relations and Political Science' (Graham 2004d, 24).

In the past fifteen years, however, urban geopolitics, an interdisciplinary and diverse corpus of academic literature in urban studies and architecture, geography and political science, has engaged with the role and meaning of cities within the discourses, practices and spaces of global politics. Urban geopolitics was first mentioned in Francophone scholarship within the context of urban power struggles in Quebec's cities (Hulbert 1989) and then, in the English one, in investigations of urban military operations in Latin America (Demarest 1995). We can define urban geopolitics as the study of the urban space of global politics. Urban geopolitics relies on a common interest in tracing the specific connections between urban spaces and places, localized dynamics of power and violence, and wider discourses and practices of global politics (Amar 2009; Douzet 2001; Fregonese 2012b; A. Ramadan 2009; Yacobi and Pullan 2014). The connections between cities and geopolitics have also tangentially concerned critical urban and conflict studies (Coaffee 2003; Coward 2009; Morrissey and Gaffikin 2006; Pullan 2011; Yiftachel 1998; Yiftachel and Yacobi 2003) and science and technology studies (Brand and Fregonese 2013), producing an eclectic body of work that is very recently expanding its conceptual and empirical repertoire (Rokem and Boano 2017). It is worth reviewing some of these contributions to conceptualize and contextualize the conflict in Beirut within a broader view of cities and their role in global politics, and of the importance of urban space and the built environment in shaping dynamics of political violence and understandings and practices of sovereignty. The following chapter reviews: first, the contribution of urban geopolitics for the study of cities, war and discourses and practices of statecraft. Second, it presents the idea of *urbicide* intended as the wanton violence against the city's built fabric and social life. It then delves further into the ideas of urbanity that are targeted by urbicide, and finally develops further the notion of urbicide borrowing from within geographical theory.

## Contested histories

With the majority of the world's population living in cities, conflict and politicized violence nowadays have become paramount urban issues. Warfare globally is also becoming asymmetric (Kaldor 1999), as many conflicts are fought between actors with substantially different levels of tactical and strategic capability, equipment and powers – usually seeing professional state armies engaged in open warfare and guerrilla against insurgencies and irregular armed groups.[1] These changes imply 'the unraveling of the national model of citizenship and service at the center of geopolitical forms of state and population security' (Cowen and Smith 2009, 36) and a reterritorialization of violence where non-state actors like militias, drug cartels and terrorist cells are involved in conflict against states and produce spaces and practices of power whose grounded mechanisms remain understudied. Historically, war violence and deliberate attack, siege and destruction have always been part of the history of cities and city-states. The construction and the destruction of cities are reciprocally intertwined as 'cities, warfare, and organized political violence have always been mutual constructions' (Graham 2004b, 1).

Cities haven't always been this important geopolitically, though. While city-states and free cities punctuated Europe until the end of the nineteenth century, 'modernity transformed the relationship between cities and states' (Shaw 2004, 142). Cities would no longer be the defining nodes of power and defence, but rather absorbed in and contained within the territorial borders of the nation state. Between the sixteenth and the nineteenth centuries, political violence together with culture, identity, and territory was gradually framed within the legal and semantic realm of the modern nation state across Europe, and although the city continued to constitute a hub for production and administration within the state, they 'were no longer the organizers of their own armies and defences' (Graham 2004b, 2). However, since the nation state became the legitimate protector of citizens within its territorial borders, and defensive city walls were metaphorically and often physically torn down (Aibar and Bijker 1997), violence in and against cities was removed from urban social science (Bishop and Clancey 2004) – and, as we will see in Chapter 5, from the rhetoric of statecraft and official foreign policy. Security was relabelled as something concerning almost exclusively state borders and armies; citizenship coincided with the territorial unit of the nation state, as 'increasingly […] the "imagined community" of the nation state, subsume[d] the city as a motive force in history' (Soja 2000, 24). Despite these depictions, war, political violence and politicized planning have indeed shaped cities throughout modernity, the twentieth century and the Cold War (Farish 2004). Contrary to the Weberian view of cities as loci of civility (Isin 2002) and to the belief that planning implicitly tends towards (peaceful) progress, Graham (2004a, 52) argued that there is a '"dark" side of urban modernity' that cannot be ignored any longer.

According to Bishop and Clancey (2004) from the First World War and until the Vietnam War, there has been a reluctance, in Western (and especially United States) cultural production and rhetoric of war, to engage with the bodily and grounded aspects of violence in and against cities. Not only cultural narratives and poetics of war, violence and death mainly relied on disembodied tropes like homeland, honour or the unknown soldier, but the economy of images around violence and death in cities was also one where bodies, blood and the technologies used to wreck havoc in cities like Hiroshima, Tokyo and Nagasaki through firebombing and atomic bombing during the Second World War were evoked in disembodied ways: 'Hollywood [...] never portrayed the actual bodily horrors of nuclear warfare' (Bishop and Clancey 2004, 61). Or they were even turned into repeatable and commodified images: 'So passionless, disembodied, and consumable was the [Hiroshima] mushroom cloud image that it became the icon on many American consumer products in the middle to late 1940s, helping flog everything from toothpaste, drive-in movies, [...] to special drinks at bars.' (Bishop and Clancey 2004, 60). Urban studies and social sciences, furthermore, have not engaged enough with the critical analysis of urban spaces as the physical targets of violence because of 'the almost complete dominance of national rather than sub-national spaces and politics within international relations and political science' (Graham 2004d, 24). Contributing to the rhetorical removal of death from the city was the association of urban disaster with the colonial city, as opposed to the orderly European modern metropolis between the end of the nineteenth century and the beginning of the twentieth century. Modernity in Europe coincided with the assertion of the security and integrity of imperial metropoles from what Bishop and Clancey (2004) term 'urban drama': 'the demarcation between modern and ancient, from the perspective of the Nineteenth century, between the time when whole cities were destroyed and their inhabitants slaughtered, and the time when that no longer happened' (Bishop and Clancey 2004, 56). This was possible because the imaginary of urban drama had shifted from the metropolitan centres of Europe to its colonial towns, in the form of either natural or anthropogenic socio-environmental disasters (such as destructive earthquakes and famines), or even wanton violent attacks. All this could take place 'in a non-European world read as still ancient and/or subject to rule by Nature (including human natures in need of taming)' (Bishop and Clancey 2004, 56). As Lindqvist (Lindqvist 2002) has illustrated, from Beirut to Damascus, from Kabul to Tetuan, Tripoli and the town of Somaliland and Darfur, sweeping aerial bombardment by European armies aimed at subjugating the local populations, going beyond the distinctions between civilians and warriors decreed by the rules of international law of La Hague (1907).2 Destruction in the colonial city was often explained through the rhetoric of the civilization mission of Europe towards 'the Orient'. The colonial city was not only bombed to rubble, but it was also *planned to rubble*. Demolition and restructuring of the previous fabric of Europe's colonial cities such as Algiers, Marrakech, Cairo and indeed Beirut were rationalized through discourses of sanitation (Lisle 2016), beautification and modernity, with spatial planning ad architecture as spatial implementers of the colonial rule (Çelik 1997; Mitchell 1988a).

These discourses were enacted in Beirut by the municipal authorities between the last years of the Ottoman Empire and the proclamation of the French mandate in 1920, including new paved roads and railways to connect Beirut to its hinterland, the construction of new governmental buildings, markets, hospitals, schools and other projects regarding 'health, embellishment, and traffic control [which] endowed the city with "modern" and "civilized" attributes, i.e. Western features' (Yacoub 2010, 7). On the basis of the new municipal building laws issued in 1896, Beirut's Ottoman Governor Azmi Bey decided to start a new turn of modernization works for Beirut in 1915. Besides demolishing the old markets or 'souks' (a ceremony for the demolition was staged on 8 April 1915) and issuing three days' notice warnings of evacuation before demolishing whole portions of the old city (Yacoub 2010), Azmy Bey had another large avenue carved in the middle of the old city, and when the French allied troops disembarked in Beirut just three years later at the end of the First World War and just one year before declaring their colonial mandate on Lebanon, they 'found a large breach in the city-centre consequence of the modernization projects. They will immediately proceed with the destructions clearing the old part of the city that was still preserved' (Yacoub 2010, 26), and with the renaming of streets transforming the Ottoman topography into a French one, preparing colonial Beirut to host the world fair in 1921, where the Beirut pavilion, despite the eclectic new architectures of the surroundings, preserved a neo-Moorish style (Yacoub 2010). The modernization of the urban fabric of Beirut, culminating with the hosting of the world fair in 1921 and involving an intense reworking of the extant plan of the city, reflected the Cartesian principle of representation of the 'world as an exhibition' that was applied to the societies and to the built fabric of the colonial 'Orient' from Algiers to Marrakech and from Cairo to Delhi between the end of the eighteenth and the twentieth centuries:

> The colonial city was to be constructed, like a world exhibition, as a representation set up before the mind of an observing subject. The Cartesian mind was conceived, in a similar way, as an interior space in which representations of external reality are inspected by an internal eye – in other words, again, like an exhibition set up before an observer. (Mitchell 1988a, 177)

The colonial destroyed city – by air bombing or by planning – shaped many Eurocentric representations of 'Oriental' lands. This involved an epistemological division that is still with us (Graham 2006; Gregory 2004, 2006) between our cities – ordered, intact and to protect as such – and their cities – backward, spatially confusing and even prone to 'natural disaster against geographies already considered disordered, violent, and overly spontaneous' (Bishop and Clancey 2004: 56). This threshold between the Western 'intact' city and the Oriental 'destroyable' city was unsettled during the Second World War when the bombing of European cities (especially in Britain and Germany) and Japanese ones involved vast scale area bombing (Hewitt 1983). In 1983, Ken Hewitt urged scholars to explore the 'terra incognita' (258) of the socio-cultural and technological connotations in the urban destruction of the Second World War. Death occurred in the European and Japanese cities, but it was very rarely depicted as urban: the nation state army and the homeland were rhetorically portrayed as the main victims and even the vision of the dead body of the soldier was absorbed into a wider, disembodied idea of homeland (Bishop and Clancey 2004). Furthermore, the detached bird's eye perspective of the bombardier deprived its targets of corporeality (Sebald and Bell 2004). The atomic death by 'vaporization' suggested a blood-free image of death to the American public at the news of the bombings of Hiroshima and Nagasaki, which were officially portrayed as a pre-emptive action to protect US cities from a Japanese invasion, added to a generalized absence of information among the American public about what happened on the ground in Hiroshima and Nagasaki until the 1980s (Linqvist 2003) has contributed to further void urban death of corporeality.

The removal of death has continued into the Cold War, especially in the United States. Farish (2004) has analysed the links between the politics and rhetoric of atomic danger and town planning in 1950s US suburbia. A discursive (and physical) demarcation was constituted, by public officials and planning alike, between inner cities and the new lively suburbias through depictions of the former as degraded, dangerous, and promiscuous, as well as by encouraging decentralization (Farish 2004). Planners conceived new anti-bomb suburban neighbourhoods, satellite cities that were self-sufficient and organized to stand a nuclear attack, and above all separated from the inner city by green belts (Farish 2004). While the focus on decentralization steered attention away from the possibility of urban death from the American public during the Cold War, Farish also remarks the political and strategic value of these apparently a-political spatial planning devices. Justifying decentralization with the perils of inner-city life, urban cores were thus 'left behind by the combinations of geopolitics and science during the Cold War' (Farish 2004: 109) not only to decay, but also to get hit by and absorb the bomb, and thus shield suburbia from the spread of the atomic plague.

## New urban wars?

Recently, cities are figuring again in geopolitical reasonings. The rhetorical and military targeting of cities are seen as a feature of the specific geopolitical moment we live in the post–Cold War phase, and are often situated within discourses about the global loss of power of the nation state (Graham 2004c). This approach is often found in studies in international relations and political science about the appearance of 'new wars' (Kaldor 1999). The 'new wars' discourse describes the military and territorial novelty of the conflicts that have developed since the end of the Second World War and especially after the end of the Cold War. According to these studies, totally new theoretical categories and policy strategies are required to understand, manage and prevent these types of war. Globalization and its impact on the decline of nation state power are deemed to be the main causes for the emergence of 'new' asymmetrical wars that involve a range of actors other than the nation state such as militias, drug cartels, terrorist cells and many others (Graham 2004d). However, critical voices have scrutinized this novelty claim (Brzoska 2004; Henderson and Singer 2002), but too often use quantitative and structural approaches which tend to focus on the differences between categories such as 'civil wars', 'inter-communal wars' and so on (Gurr 1994) rather than on the particular spaces these so-called new wars produce. On the other hand, urban geopolitics has also put an emphasis (albeit qualitative) distinction of post–Cold War conflicts and especially on urbicide as the feature of these wars (Coward 2006). One of the main ideas underpinning studies in urban geopolitics is that with the supposed de-territorialization of the power of the nation state in a global world, and with the increasingly complex transnational flows putting the coherence of territorial and national discourse into question, cities are becoming the new centres of the rescaling of political violence.

This rescaling is only one of the aspects of a broader 'repositioning' of notions of identity, territoriality, citizenship and locality in an era of globalization. For Sassen (2005) these state discourses and practices of power are not purely de-centred and losing their concrete specifications in a flat world of flows, but on the contrary, they are 'repositioning' and creating new spaces of political enactment. In the global world 'The national as container of social process and power is cracked. This cracked casing opens up possibilities for a geography of politics that links subnational spaces. Cities are foremost in this new geography' (Sassen 2005, 89). The city in the global world, therefore, becomes 'one of the nexuses where the formation of new claims materializes and assumes concrete forms' (Sassen 2005, 89) after the nation state and concepts such as 'homeland', had constituted – especially during the two world wars – the categories through which state governments made sense of conflict and represented the geographies of military attack.

The decreasing power of the territorial nation state corresponds to the increasing visibility and strategic value of what Michael Sorkin has recently called 'the city after Clausewitz' (Sorkin and Graham 2004, 262), where 'the era of the suicide bomber' (262) produces discourses and policies of daily prevention, fear and deterrence, against complex actors of political violence – 'drug entrepreneurs, jihadists, local liberation fronts, animal right activists, abortion abolitionist, and the rest' (262) – whose claims and spaces of power have very little to share with national territoriality in a classic sense.

In military studies, cities have been considered for the past two decades at least as the renewed crucial terrains of future wars (Press 1998). Such statements contribute to form global geopolitical imaginations where cities are natural arenas for conflict. There is a risk here of operating a mere rescaling of conflict from the national to the urban level, which can have the consequence of normalizing urban war as a default condition of the post–Cold War world. As Kassir's words at the beginning of this introduction hint to, and what this book aims to demonstrate, is what Graham – drawing on Appadurai (1996) – has called 'the implosion of global and national politics into the urban' (2004c, 7), which was present in Beirut well before the end of the Cold War and from the very first weeks of fighting in 1975. I want to highlight the profoundly urban character of war in Lebanon in order to widen the perspective of urban geopolitics beyond post–Cold War conflicts and to explore how regional and global geopolitics played out very powerfully at the urban level even before the era of 'new wars'.

Until recently, urban geopolitics' has had an analytical over-reliance on few, heavily militarized case studies, for example, Israel/Palestine (Graham 2005, 2004e). Relatedly, Arab cities feature recurrently in urban geopolitical reflections as the orientalized target grounds of the War on Terror (Gregory 2006) or as agglomerations of 'complex' infrastructures and insurgencies that obstacle the US technomilitary dominance (Graham 2008). This technocentric viewpoint overlooks the importance of ordinary urban sites, that are not necessarily the derivation of militarized warfare, not only in becoming targets but also being urban mediators of wider geopolitical processes. Urban geopolitics, then, ought to pay more attention to the 'normal' sites that do not derive from warfare (Harris 2015), and to the 'ordinary topologies' of urban contestation (Harker 2014). Without diminishing the importance of militarized conflict settings, urban geopolitics needs to shed light also on the micro-practices and discourses underpinning non-state political violence *in*, *against* and *through* cities such as that by armed irregular militias. Despite the existence of quantitative analysis of asymmetric conflict and non-state actors in conflict (Cunningham, Gleditsch, and Salehyan 2013; Sundberg, Eck, and Kreutz 2012), more qualitative and empirically driven research is needed to identify the grounded mechanisms of non-state violence (Fregonese 2015), and that 'the literature is still in need of a better understanding of the microgeographies of small wars' (Korf 2011, 733).

A second, more conceptual critique has been moved in the past decade to urban geopolitics by political geography scholarship (Flint 2006; Smith 2006). The question of whether urban geopolitics is, after all, just a rescaling of geopolitics (and associated practices of war and statecraft) from the national to the urban. This critique warns against simply replacing states with cities as the new necessary terrains to for interpreting and acting in global politics, hence as the normal and loci of war, rather than producing deeper spatial understandings of urban conflict and the role of cities in contemporary global challenges (Bialasiewicz 2015). In other words, this point has left some scholars wondering 'why we are heading towards an urban geopolitics and what that may be when we arrive, and how it will be different from an undefined initial geopolitics' (Flint 2006, 217). This, however, was also a concern of earlier urban geopolitics work, warning against the perils of de-politicizing and institutionalizing knowledge about city warfare (and the violence suffered by civilians in cities) into 'a technoscientific discipline with its own conference series, research centres, and journals' (Graham 2005, 1).

A third critique tackles urban geopolitics' techno-centrism and its production of a disembodied approach to urban violence, which does not address the spaces of lived experience and feeling bodies (Harker 2014). Here, Adey (2013) calls for complementing spatial accounts of urban militarization with less representational (and techno-centric) perspectives about urban security and militarism that highlight how they 'are experienced and made present to the lives that live them' (Adey 2013, 52), in order to go 'beyond the scientific and geological political-technological imaginations deployed in the bundles of the strategic and the technical' (Adey 2013, 52).

My investigation of the urban geopolitics of the Two Years' War in Beirut tackles the tension between, on the one hand, contextual understandings of the impact of war on cities and, on the other hand, the tendency to normalize cities as default battlegrounds of contemporary conflict. To clarify this tension further, we ought first to reflect on the (geo)political value of the material fabric of cities and the role of the urban built environment in political violence.

# Urbicide: Theorizing political violence against the urban built fabric

Urban geopolitics literature generally stresses how one of the features of current warfare is the increasing presence of wanton violence, against the built environment of cities (Gregory and Pred, 2007), or 'urbicide', the 'deliberate killing, of the city' (Graham 2004c, 25). Urbicide is described as one of the defining features of post–Cold War conflicts. Political scientist Martin Coward (2009) especially notes how the destruction of the built environment in intra-state conflict 'seem[s] to represent a particularly vicious, form of warfare that had novel distinguishing features' (2009, 37), and that we 'must take into account the assault on buildings, logistics networks and communications infrastructure' (2009, 121) in order to understand conflict in the post–Cold War world. This violence, I argue, is the product of historically specific and contextually charged political struggles over space, rather than simply a result of wider geopolitical shifts such as the end of the Cold War and the rise of asymmetric conflict characterizing the so-called new wars (Kaldor 1999) and the urbanization of warfare (Sassen 2010, 2018).

## A genealogy

Destruction and construction of the urban socio-material fabric very often proceed in parallel during processes of modernization (Aibar and Bijker 1997), capital accumulation (Harvey 2008) and gentrification (Berman 1984), as well as political violence and deliberate targeting. These studies view the city as producing new relations out of violence and destruction. The term 'urbicide' (city-killing) was first used to describe the impact on the built environment and the communities living in it of the 1960s aggressive urban restructuring policies in the United States (Berman 1984; Huxtable 1989). Berman used the idea of urbicide to highlight the violence implied in dynamics of urban regeneration in South Bronx following the planning by Robert Moses of an expressway cutting through neighbourhoods with consequent wide socio-economical unsettlement. Berman even called 'indirect displacement' (1996, 174), the practice by developers of acquiring land and letting buildings decay to the point of making them uninhabitable.

'Urbicide' was then later reappropriated by reflections around armed conflict, to indicate attacks against cities during the Balkan wars (Bogdanovic 1994; Chaslin 1997; Mostar Architects Association 1993) and, specifically, to indicate the 'revenge of the countryside' against the multicultural cities and their material fabric (Bougarel 1999; Ramet 1996; Simmons 2001). At the end of 1991

Bogdan Bogdanović, the former mayor of Belgrade, described as 'urbicide' the deliberate destruction of the built environment in the actions of the Yugoslav National Army during the siege of the Croatian border town of Vukovar (Safier 2001, 422). The destruction of the built fabric of towns and cities, he argued, was part and parcel of the ethnic cleansing campaigns. He denounced the violence perpetrated both against humans and the built environment, noting that the horror of the West towards that conflict was 'understandable: for centuries [the West] has linked the concepts "city" and "civilization" […]. It therefore has no choice but to view the destruction of cities as […] opposition to the highest values of civilization' (Bogdanovic 1994, 57). During the sieges of Sarajevo and Mostar by the Yugoslav National Army in 1992, the notion of urbicide stirred up a debate among local architects about the destruction of the urban fabric in Bosnia-Herzegovina, resulting in an article titled 'Mostar '92 – urbicide' published in the magazine *Space and Society* (1993).

Urbicide has also been debated in the European academia, and was the focus of a joint colloquium in Lyon and Geneva in 2000 where 'the vulnerability of cities so far forgotten' was debated in the frame of 'new local conflicts' and new urban emergencies (Baudouï and Grichting 2005). Nearly a decade after the war in former Yugoslavia, urbicide started gaining attention also from the Anglophone academia, with attempts to conceptualize the impact of 9/11 on everyday urban life (Marcuse 2004)(Marcuse 2001), on changes in discourses and practices of security, and on understandings of multiculturalism (Leontidou 2001; Safier 2001). The symposium *Urbicide: The Killing of Cities?* was held in 2005 at Durham University, resulting in a special issue of *Theory and Event* (Volume 10, Issue 2, 2007).

Coward theorizes urbicide to overcome what he calls anthropocentric perspectives on conflict:

> The problem with [the semiotic] interpretation [of political violence against the built environment] is that the destruction is not treated as an event worthy of attention in its own right. Rather, the rubble is appropriated as a sign connotative of a more general concept. While urban destruction may serve as the sign for several concepts, noting this does not get us any closer to understanding the meaning of the destruction of urban fabric. […] We deny ourselves crucial political possibilities if we simply accept those significatory stories, since we accept that this destruction is interesting only insofar as it connotes the dissolution of political communities or the savagery of this fragmentation. (Coward 2004, 164–65)

Coward instead opens up a debate on political violence, targeting and victimization that considers objects and the built environment not only as collaterals of military operations, or symbols of coherent human identities, but as agents and indeed targets in wanton acts of political violence. The use of built forms as targets of war in themselves has been also analysed by Graham (2003) in the Israeli–Palestine case, focusing on the ways Israeli Defence Force (IDF) military operations against Palestinian occupied territories in the early 2000s involved not only symbolic and institutional buildings, but also mundane infrastructure such as apartments, offices and sewerage. Aimed (according to the IDF) at the 'eradication of terrorism', particularly through the use of bulldozers, 'Urbicide involves not just the demolition of homes … but intensive infrastructural destruction' (Graham 2003, 19:66) including roads, water tanks, power plants and communication systems. Architect Eyal Weizman (Weizman 2004) places an even deeper stress on the importance of the everyday built environment and its organization in influencing modalities, effects and management of a conflict. When considering the Israeli tactics of domination through eviction of the Palestinian populations, urban planning and management are – according to Weizman – among the primary means of achieving that goal. Weizman's (2003) original case for what has become the Forensic Architecture research agency (https://www.forensic-architecture.org/) was about bringing urbicide out of academia and into the legal realm, as part of international legislation on war crimes, and called for holding architects responsible for crimes and damage to civilians derived from architects' spatial decisions and the organization of the built environment in which they have participated.

## The city of urbicide

But what exactly is 'killed' in urbicide? What is 'the urb-' of urbicide? Bogdanović's view that 'urbicide […] is the intentional attack on the human and the inert fabric of the city with the intent of destroying the civic values embodied within it' (in Bevan 2005, 121) conveys the idea that cities have a particular 'civil' way of life defining them, and that this is embodied in their built fabric. Bogdanović, therefore, interprets political violence as the expression of hostility towards a distinctively *urban* way of life. This is a view adopted by other scholars, who identify that hostility with the countryside's view of the city as a corrupted place (Simmons 2001). It is also a view that Omar, one of the former combatants who participated in the research, expressed. Omar views Beirut as a city that, between independence in 1943 and the civil war in 1975, became a modern capital. But with the civil war, it lost its character as a city and retreated instead into what Omar calls 'a rural mentality'. The culture of the city, Omar continues, is one of acceptance of the other, which is different from rural life in a village.

However, this view of a confrontation between an introverted countryside and a cosmopolitan city romanticizes and 'perpetuates a modernist myth' that considers city and countryside as two distinct containers of culture, identity and practices, where 'the city represents progress whilst the countryside remains backward' (Coward 2006: 425). This binary isolates these spaces from their contexts of social and material production, from their interaction and mutual constitution; and sweeps away many of the unique and contextual aspects that every conflict has. The urban versus rural trope is inadequate to explain the urban dynamics of the Two Years' War in Beirut. Beirut's experience of conflict is neither the result of the actions of a rural mass that was hell-bent on the destruction of the city nor the opposition of cosmopolitanism versus sectarian chauvinism. In the history of political violence of Lebanon, the 'rural versus urban' thesis does not hold, for a number of reasons. First, Lebanese cities have not always been the prime objects of attack: the centre of the sectarian clashes in the mountainous areas of late Ottoman Lebanon in the 1840s and 1860s were rural villages, where farms, fields, vineyards, religious buildings and private residences were targets of destruction and looting (Makdisi 2000). Second, attacks against property and material damage – including against religious buildings[3] – took place in rural areas, as well as in cities, throughout the civil war (Labaky and Abou Rjeily 1993). In essence, the historical geography of violence in Lebanon from the nineteenth century onwards does not map simply onto an urban/rural divide.

The vision of urban place as a container of distinct culture is reinforced by architecture historian Nicholas Adams who wrote, in 1993, about the role of architecture as a target of deliberate violence. Adams interpreted the targeting of architecture as the attempt to eradicate the presence of certain people from certain places, the former seen as belonging to an 'alien culture' (Adams 1993, 389), and the places where they gather and conduct their public life become therefore targets for hostile actors. Both Adams and Bogdanovic's views of the correlation between violence and urban material fabric are problematic. They both view the city – especially the Western city – as a discrete container of civility. For Bogdanovic, the attack to the urban built fabric an attack to some 'values' embedded in urban space and are celebrated in the civil 'way of life'.

From a spatial viewpoint, Isin (2002) has extensively deconstructed the Weberian notion of city as a space of civility and association of civil subjects. His history of accepted ideas of citizenship analysed how 'being political' is in reality a 'becoming political': it is a process of reciprocal constitution of self and other through 'acts' within the lived (public) space of the city. In other words, he highlighted the power contest at the basis of the process of 'becoming political' (4) in cities. There is no political being that exists as already constituted 'citizen' outside the lived and contested space of the city: 'Being political means being of the city. There is no political being outside the machine' (2002, 284). What Isin means is that political subjects (those subjects who have right to the city) are socially constructed not in isolation, but in relation to those who are not political, that is, not of the city (such as, in Isin's account of ancient Greece, outsiders, slaves, and women). Isin thus shakes the foundation of the romanticized notion of the city as a 'harmonious and contiguous' (2002, 3) association of citizens, by tracing instead the social contexts in which citizenship is defined (with all sorts of exclusions and inclusions). The result is a proposition for a new view of the city, not as an association of subjects who are already citizens/civil, but as the association of those subjects who have become citizens by struggling for the determination of what constitutes the uncivil.

Bogdanovic's view of urbicide is one of an attack to an already constituted and immobile idea of city/civility and of equally immobile non-civil subjects. His view reflects Weber's romanticized idea of city as association. Instead, Isin presents the idea of city as a machine of differentiation, where political subjects, rather than 'being' civilized a priori, *have become* citizens through differentiating themselves from the 'Other' which they constituted as external, inferior and different through contested and contextual practices. In Bogdanovic's reasoning, the built fabric that is made target is a referent of supposedly coherent values of civility that the non-urban people want to cancel. This reasoning, however, deprives the built environment of its contextually shaped nature and the conflict of its specific agencies. Here, the built environment is only attached to what are already predetermined and conclusive ideals, but is not constitutive of those ideals.

In Isin's city of differentiation, instead, the urban built fabric is neither a merely physical background for the struggle for affirmation of political subjects, nor a purely abstract and already composed value to struggle for or against. On the contrary, the built environment 'becomes an object in the struggle for domination and differentiation' (Isin 2002, 49), the element *through* which a group differentiates itself. We could then say that the built environment is a constitutive agent of what is defined as citizenship/civility/city, rather than a background or an ideal superstructure. The city therefore, for Isin, is 'never simply a passive background of becoming political. It is a fundamental strategic property by which groups, nations, societies, federations, empires, and kingdoms are constituted in the real world' (2002, 49).

In the words of George Corm, during the civil war in Beirut the militias were 'the true occupants of the country [...] they controlled every square centimetre of territory' (Corm 2005, 204). Engaging with the geographies of the fighting on the ground, and exploring both the everyday practices and representations that militia fighting produced out of a close-quarter war in a dense urban environment, offers the possibility of a richer discussion of urbicide, complicating superficial explanations of violence against the city as a manifestation of urban/rural rivalries, as well as adding context to its generalization as a feature of post–Cold War conflicts everywhere.

Coward has used the philosophical approach of Martin Heidegger to conceptualize urbicide while shifting away from semiotic/symbolic interpretations of the built environment as a container of culture and advancing instead a theory of urbicide that is more grounded and appreciative of contextuality. Coward observes that during the war in the former Yugoslavia, what was targeted were not only emblematic or significant buildings (such as churches, mosques, and monuments) symbolically connected to one or another victimized community, but also mundane edifices such as shops, bridges, homes, and even post offices. What was attacked then, according to Coward, was not only the highly symbolic urban fabric that one could identify with wider ideas of 'nation', 'religion', or 'identity', but also any other spaces where diverse people would conduct mundane activities, and through those spaces (and the activities they enabled) connected in everyday life. Urbicide then, argues Coward, can be seen as the violent attempt to erase the conditions (and spaces) of urbanity, conceived as heterogeneity and – drawing from Heidegger – a condition of 'dwelling together' (Coward 2004). The urban fabric, then, becomes a target of violence where it represents and enables coexistence among different communities:

> Insofar as the dynamic of ethnic cleansing is that of the carving out of separate, ethnically homogeneous and self-determining territorial entities, it comprises a denial of common space through a destruction of that which attests to a record of sharing spaces – the heterogeneity of cultural heritage and the intermingling of civilian bodies. (Coward 2004, 158)

Despite Coward's tendency to situate urbicide as a feature of post–Cold War conflicts that with this book I wish move beyond, Coward's considerations on heterogeneity resonate particularly in the context of Beirut. In Beirut as in Coward's Bosnia, the 'politics of urbicide' (2004, 167) have as objective deleting the material possibility of encounter between different communities. The politics of urbicide are aimed at keeping people separate, both physically and through the representation of 'us' and 'them', a mechanism that is sustained by 'the fiction of ethnic [or sectarian, in Beirut's case] separateness/purity' (170). Here, as we will see, the built environment is a crucial factor that actively shapes the divisive politics of conflict, rather than being an inert background on which events simply unfold. In the next chapter, I will offer a historical geography of those divisive politics and their connections to modern colonialism, in order to understand Lebanon's more recent urban geographies of violence.

2

# Modernity, Territory and Conflict in Lebanon

The fifth stage, which lasted from the Nineteenth century to the establishment of Greater Lebanon, was the colonial period, in which the Western powers used al-ta'ifiyya[1] to promote their own interests. In the early twentieth century, the sixth stage, 'al-ta'ifiyya' became part of the history of the country, [and] has become real, [though it] remains hidden from sight behind [the] disguise [of such] twentieth-century concepts such as nationalism, scientism and democracy. *(A. Sayigh 1955 cited in Firro 2003)*

## The sect as laboratory of modernity

The political geography of the portion of Mediterranean Levant corresponding to contemporary Lebanon has experienced profound changes between the Ottoman-Egyptian war in the 1830s (with the consequent demise of the feudal Shehab Emirate) and the aftermath of the First World War in the 1920s. Before the birth of the nation state of Greater Lebanon under French mandate in September 1920, the term *Jabal Lubnan* (henceforth, Mount Lebanon) was used to indicate the mountain range behind the coastal towns of Beirut and Sidon which, until the 1830s, belonged to the Ottoman provinces of Syria (*Sham*) and Tripoli (*Trablussham*). Between 1840 and 1860, the region of Mount Lebanon became *Cebel-i Lübnan Da İki Kaymakamlığı* indicating, in Ottoman Turkish, the two Lieutenancies of Mount Lebanon (Henceforth: Double Kaimakamate), one Druze and one Maronite. In 1860, the autonomous province (*Mutasarrifiyya*) of Mount Lebanon (*Petit Liban*) replaced the Double Kaimakamate and extended to the coast. In 1920, the nation state of Lebanon under French mandate resulted from the incorporation into the Mutasarrifiyya of the Vilayats of Tripoli, Sidon and part of the province of Damascus, thus establishing the boundaries of present Lebanon.

Defining the spatial extent of modern Lebanon is complex but crucial for this book. Some authors (Salibi 1976) distinguish (modern) mandate Lebanon from Ottoman Lebanon on the basis that only with French colonialism did Lebanon acquire the political structures of a modern nation state. Others focus, more interestingly for this book, on the previous phases, starting with the administrative reforms in the Ottoman Empire in the 1840s. These scholars consider the modernization of Lebanon as a nuanced and contested and even at times contradictory process, made of complex influences between Ottoman authorities, European powers, local elites and the wider population. In this process, the religious community or 'sect' (in Arabic: taifa) emerged as a key dispositif[2] through which European colonial powers and Ottoman authorities governed the population: counting and mapping it, partitioning to pacify it, eventually embedding the taifa within the executive, legislative and judiciary fabric of the emerging nation state, and deeply changing local discourses and practices of subjectivity, identity, history and territory.

In this chapter, I aim to provide a spatial account of the emergence and affirmation of sectarianism (ta'ifiyya) as a territorial and governmental instrument in Ottoman, colonial and independent Lebanon, initially serving as criterion for partitioning Mount Lebanon, and eventually becoming embedded in the governmental structure of the territorial nation state. The spatial genealogy of Lebanon's political sectarianism as a modern tool of governmentality in this chapter sets the complex historical background for the disputes around sovereignty, the Lebanese question and physical partition that reshaped the urban geography of Beirut during the Two Years' War.

The following pages place the political violence of 1975–1976 within a wider historical frame of recurrent localized violence in late-Ottoman Mount Lebanon, and trace the vast geopolitical connections of these events, reaching well beyond the confines of the towns and region where they took place. This historical perspective serves to complicate accounts of the civil war, held by the press and most importantly by international official circles in 1975 and 1976, as an essentially religious clash of ancestral communal rivalries between Muslim, Christians and Druze, where sectarian leaders and their militias were driven by their own religion to annihilate their opponents. The spatial account in this chapter instead provides a view of the taifa not as an age-old identity, but as the result of specific geopolitical and spatial projects advanced by European powers and by Ottoman authorities at a time when the traditional social structures of this part of the Ottoman empire encountered the epistemological and spatial project of colonialism.

Scholars have extensively argued that these complex, multiple and sometimes contradictory encounters of late-Ottoman Mount Lebanon with colonialism resulted in a gradual discursive formation and establishment of sectarianism as a political and governing tool (Tarazi-Fawaz 1994; Firro 2006; Hakim 2013; U. Makdisi 2000). However, missing from these reflections is the territorial dimension. Instead, I argue that cartography and the representation of territory were part and parcel of the cultural and discursive production of political sectarianism at the dawn of modern Lebanon (U. Makdisi 2000), and that specific, successive and often contrasting cartographies contributed not only to shape the sect as the basis of political subjectivity and political legitimacy, but also to establish it as a fact on the ground.

Political violence between the 1840s and 1860s had specific impacts beyond the confines of Mount Lebanon, bringing dramatic shifts in the urban geopolitics of nearby Beirut. While towns in Mount Lebanon like Deir Al Qamar had, until the demise of the Shehabi Emirate, been the demographic commercial and political nodes of this part of the Ottoman Empire, from the 1840s Beirut began a process of expansion that will project it into becoming the capital of a nation state. In his study of the movements of population from the Mountain and the geographical impact on Beirut, Davie goes all the way to argue that the Damascus Road, coinciding with the Green Line dividing Beirut into two sectors (East and West) during the civil war, is not purely the serendipitous result of the fighting tactics in 1975 and 1976, but of spatial and demographic genealogies dating back 150 years to the epoch of the violence and the socio-economic changes in Mount Lebanon. Davie's argument, however, has a fundamental lacuna: it considers the state as failing to contain the sect, as a separate entity from what he describes as the population's 'original allegiances' (Davie 1992, 14). This primordialist view of the sect overlooks its role in the very construction of the nation state and in establishing and normalizing political subjectivities, legitimacy and institutions. It is important, for this chapter and the following ones, to establish that the sect is not a failure of the state, but part of the process of state formation and, equally, key to understand the battles over contested meaning of nation, territory and sovereignty – battles that have resurfaced on the urban ground of Beirut during the 1975–1976 civil war.

There are two major assumptions when it comes to relating modern Lebanon and sectarianism: firstly, that the religious sect is a given, ageless characteristic of Lebanese society and an a priori entity that is abstract from the social process. This assumption nurtures a second one which represents the sect as a pre-modern – or anti-modern – entity. Leila Tarazi-Fawaz (1983) has unpacked the first assumption about pre-existent religious identities by pointing out how specific, embodied processes – such as the mechanisms of punishment and change in social prestige that intra-faith conversions in nineteenth-century Beirut triggered – have been obfuscated by generalizations about an over-represented Christian-Druze-Muslim ageless rivalry. Fawaz distinguishes between religious violence (between the main religions in Lebanon) and sectarian violence (i.e. between members of different credos in the same religion, such as Maronite and Greek Orthodox within Christianity) and remarks that the stress put on a pre-given, abstract separation between the main religions misses out on the contingent and daily development of discourses about sectarian identity and, therefore, does not explain the grounded mechanisms whereby religious or sectarian violence is triggered: 'emphasizing hostilities between Christians and Muslims, or Christian and Druze, or Druze and Muslims, misses the point that, until the end of the Nineteenth century, sectarian hostility in the city was as much within as among the major communities' (Tarazi-Fawaz 1983, 108). The second assumption on sectarianism describes Lebanon's sectarian politics as un-modern and 'tribal'. Makdisi (2000) has eminently unpacked this assumption, showing how the sect instead played a decisive role in the modernization of Lebanon. Imperialism enacts a 'technology of colonialist subjectification' (Ashcroft, Griffiths, and Tiffin 1995, 426) through the creation of modern institutional spaces of knowledge and discipline. The introduction of modern ways of viewing, performing, and managing the self and the political in late-Ottoman Lebanon was very much a 'gentle crusade' (Makdisi 2000, 15) aimed at regenerating hearts and minds of the local population, not only through religious doctrine, but also by changing several aspects of everyday life from education, to inhabiting, to care. Makdisi (2000) offers examples about the introduction of modern science, medicine and other everyday modern devices in late nineteenth-century Levant by the Jesuits and other religious institutions: '(Western) furniture, education, medicine, and knowledge' (89). Secular education, by far preferred by the local notables, was still administered by the clerics, who on one hand sought to satisfy the demands of the local elites by providing laic education, but on the other hand piloted the affluence to this laic education along sectarian lines, so that, for example, Catholic and Protestant missionaries would try to attract different parts of the local communities: the Protestants focused on the Druze, while the Jesuits tried to win the hearts of the Maronite Catholic (Makdisi 2000, 90–91).

Considering these aspects allows us to reframe sectarianism as the condition for the birth of a Lebanese modern nation state, rather than as the symbol of its failure during the civil war. In the words of Makdisi, 'Sectarianism as an idea draws its meaning only within a nationalist paradigm and hence […] it belongs to our modern world' (2000, 13). Far from constituting a 'decline into barbarism', sectarianism was the political dispositif that placed Lebanon on the path of becoming a modern nation state, starting from a time in which the 'Oriental question' attracted the gaze of the Western powers toward these lands of the Mediterranean Levant.

## The refuge and the battleground

The recent cultural and political history of Lebanon, and especially its capital Beirut, has often been depicted using a double refrain of coexistence-and-contention between its religious communities (*tawā'if*). Such representations have been utilized extensively, from European Orientalist travellers' accounts to contemporary popular accounts.

Beyond the physical seizing of overseas lands by the European empires, colonialism is also a mechanism of epistemological power over the colony, inscribing new ways of seeing its space and society. In his study of the deployment of colonial technologies of discipline and control in nineteenth-century Egypt, Mitchell (1988b, ix) argues that the 'power to colonize' consists of 'the spread of a political order that inscribes in the social world a new conception of space, new forms of personhood, and a new means of manufacturing the experience of the real'. In line with this theory, Western interests towards Mount Lebanon were expressed through the 'pen and paintbrush' of Orientalist travellers, intellectuals and artists before the 'sword and the musket' of the colonial troops (Makdisi 2000, 16). Travel reports, literary and art works resulted in a constellation of imaginative and moral geographies of these spaces which, in turn, often contributed to shape ideals of Western or European civility. Cultural travel has often coexisted and meshed with the military worlds of colonial expansion (Lisle 2016; Thompson 2012).[3] The practice of the travel to the Orient and the quest for cultural knowledge by personalities like Gerard De Nerval were part of the same imperial encounter as the quest for territorial control and geographical knowledge by European armies and fact-finding missions in the Levant and North Africa since the nineteenth century. The cultural appropriation of Lebanon by the European powers was never a unilateral appropriation of the colonized by the colonizer: instead it produced new complex networks of power among the local notables, whose benevolent attitude gained the Europeans' allegiance and even protection. Western travellers in the nineteenth century often portrayed the Ottoman Levant through images of openness and multicultural encounters. These images, in many cases, were accompanied by observations of landscape features. Eighteenth- and nineteenth-century European travellers, for example, depicted Beirut as lying at the meeting point between East and West. Beirut appeared as a place of harmony for those travellers who wished to discover the origins of Europe's supposedly Christian roots. Alphonse de Lamartine, for example, described the Mount Lebanon range overlooking the city as a site of connection between East and West: 'these are the Alps under the Asian sky' (Khalidy 2003, 22), and Gerard de Nerval wrote, 'This is Europe and Asia blending into each other with soft caresses' (Khalidy 2003, 53). However, numerous accounts also represented the communities inhabiting the Ottoman Levant as discrete: Maurice Barres, just before the beginning of the First World War, described it as a 'saraband of races' (Khalidy 2003, 58) and Makdisi, more powerfully, points out that 'although Druzes and Maronite often lived in the same village, shared the same customs and owed allegiance to the same notables, they were nevertheless described separately in Western literature and [...] the various explorers and missionaries constructed [...] a discourse of Mount Lebanon's tribal characteristics, say of Druze bellicosity, which was largely self-referential' (U. Makdisi 2000, 23).

In her critical study of cosmopolitanism and economic networks in nineteenth-century Beirut, Tarazi-Fawaz (1983, 114) describes it as 'the city that had once been a refuge', a place whose growth and economic prosperity had developed in higher measure than other Mediterranean ports, mostly thanks to trade and joint ventures between local notables and merchants regardless of their religion. In her chapter on Beirut's sectarian relations from the mid-nineteenth century (1983, 116–17) Tarazi-Fawaz accounts for escalating episodes of violence and murder between Muslims and Christians in the city. Beirut, she argues, gradually became a place where sectarian tensions developed, much sharper than in other Lebanese cities, so much so that 'by the beginning of [the twentieth] century, Christian-Muslim clashes were so common that rarely did a week go by without an assassination, or a year without a riot' (115).[4] In his study of the discursive construction of sectarian political identity in late Ottoman Lebanon, Makdisi (2000) illustrates how modern colonialism produced what he defines as a socially constructed 'culture of sectarianism' through a double imaginary of Lebanon as a container of distinct and homogenous religious groups and as a land where these different communities could also peacefully co-exist. This double imaginary is also evident in Hourani's (1976) interpretation of the changing political culture of Lebanon. Hourani's theory verges around the two spatial ideologies of the mountain and of the city. Until the creation of Greater Lebanon under French mandate in 1920, according to Hourani, the Christian Maronite idea of sect as an equivalent of a nation prevailed in the mountainous regions of late Ottoman Lebanon. After 1920, however, the political culture of Beirut and of the other cities of the Lebanese coast prevailed on the mountain ideology and nurtured instead an urban imaginary of Lebanon as a country where different religions coexist in peace next to each other, affirming their presence, but accepting plurality at the same time:

> the urban idea of Lebanon was neither of a society closed against the outside world, nor of a unitary society in which smaller communities were dissolved, but something between the two: a plural society in which communities, still different on the level of inherited religious loyalties and intimate family ties, coexisted within a common framework. (Hourani 1976, 34)

This double imaginary is not unusual to Western representations of the late Ottoman Empire. It can be regarded as the result of an epistemological tension between wanting to distance the 'chaotic' oriental amalgam of races as other from Europe, and at the same time wanting to construct a controllable order out of that chaos by conceiving subjects as belonging to distinct groups. Through his analysis of the performance of cosmopolitanism in Smyrna, Georgelin (2003) shows how Smyrna's daily urban scene performed cosmopolitism and coexistence – especially in the town's markets and port – but hid rigid homogenous social hierarchies in its residential quarters, including a mounting ethno-nationalism. This tension produces the 'contradictions' of co-existence and rivalry between groups that we find also in the politico-cultural history of Lebanon and Beirut.

Colonial perspectives about the physical geography of the area of Mount Lebanon shaped a spatial discourse of refuge, especially relating to the Christian and the Druze, that portrayed them as dwelling the mountainous landscape to protect themselves from Ottoman (Muslim) violence. Modern authors have demystified this colonial image and proved that persecution and violence was in fact perpetrated by co-religionaries as well as by the Muslim, and that Christian and Druze settlement was widespread beyond the Mountain (Naeff 2018; Salibi 1988). However, this mountainous Ottoman province, peripheral and relatively autonomous from the government in Istanbul, came to be seen by the French as particular haven of Christian salvation from the hegemony of Ottoman Islam. Its natural features – slopes, peaks, and deep ravines – were viewed as an exceptional space onto which the European cultural imaginary could project its liberal or even revolutionary ideals. The autonomy of these mainly non-Muslim, mountainous lands from their Ottoman surroundings contributed to the European representations of the Druze and Maronite inhabitants as non-Muslims prevented from flourishing by Ottoman obscurantism and therefore needing to be brought onto the path to modernity through rational, enlightened and republican ideals (Makdisi 2000). Conversely, Mount Lebanon was also viewed as a refuge for anti-revolutionary restoration: French missionaries and aristocrats, for example, saw Mount Lebanon as a mountain shelter from revolution, Judaism, freemasonry and Islam (Makdisi 2000).

The colonizing power of imperialism works also through turning 'space into time' (Agnew 1995a) and obviate the 'past greatness and present backwardness' (Mitchell 1988a, 169) by establishing and naturalizing representations of a linear path of history and modernity that the colony was bound to follow. In Mount Lebanon, the religious community and its territorial representation became a crucial factor in realizing this modern episteme. In a speech given in 1845 before the House of Commons about the political violence in Mount Lebanon, former British Prime Minister Lord Palmerston declared that

> When such men [sic] are intermixed, as they are in Lebanon, occupying the same village, dwelling on the same land, constantly meeting in the same town, it evidently requires a vigorous hand, a powerful head, a strong, determined will, together with sound judgment, to repress the tendency to disorder which must exist in such a state of society. (Quoted in Makdisi 2000, 78)

Imperialism brings an epistemological order to the colony, an order that 'appear[s] to overcome internal difference, and set up the different as something outside' (Mitchell 1988a, 171). Differentiation, according to Godlewska (1994) is what imperialist geo-power tries to eliminate in favour of cartographies providing a comprehensive visual (and epistemological) grasp of the colony:

> The critical link between imperialism and modernization lay in the argument that what was local or regional or indigenous was demonstrably inferior, and indeed an unacceptable obstacle to the national and supra-national uniformity that was the primary benefit of imperialism. Diversity was associated with degeneracy and uniformity with civilization. (Godlewska 1994, 50)

In nineteenth-century Mount Lebanon, the emergence of sectarian spaces was the product of different and contested interpretations of modernity. In particular, the local elites were gradually changing their vision and representation of political and social legitimacy from notability and social prestige to religious belonging. A new geographical imagination for Ottoman

Lebanon was therefore being gradually constituted where, in the long run, religious groups would shape political identity.

The attempt to erase the 'ambiguity' of the intermixing through the geographical partition of the religious communities in Mount Lebanon, however, was first and foremost a spatial one. What I focus upon in the next pages is how the spatial and administrative rearrangement of the region was 'enshrined for the next century and a half (up until the present time) the principle of government by sectarian appointment' (Farah 2000, xxi). In other words, between 1842 and 1860, sect became synonym with territory, and the Double Kaymakamate of Mount Lebanon (*Cebel-i Lübnan Da İki Kaymakamlığı*) was the concrete expression of this equivalence.

## The sect as territory. The Double Kaymakamate of Mount Lebanon (1842–1860)

The Ottoman-Egyptian war of 1831–1840 had seen Mehmet Ali, the Ottoman vassal in Egypt, rebel against the Sublime Porte and organize a campaign, led by his son Ibrahim Pasha, into Syria and Mount Lebanon. Here, he found an ally in Emir Bashir Shehab II, in turn supported by France. The invasion ended with the intervention of the British and Austrian fleets in support of the Ottoman Sultan and Ibrahim Pasha was made to withdraw and accept the boundaries ratified in the Convention of London[5] (Figure 1). By this point, the brief Egyptian reign in Syria had brought about a vast reshuffle of privileges and left a power vacuum that made the area prone to political and communal unrest.

On the Ottoman Sultan's side, this situation prompted a reorganization of this and other parts of the Ottoman Empire, which was deemed by European powers to be administratively and governmentally inefficient and losing control on its provinces, including Mount Lebanon.[6] The *Tanzimat* (reorganization) of the Ottoman empire between 1839 and 1876 are contained in the Gülhane Decree sanctioned by Sultan Abdülmecid I in 1839 and consist of a centralized restructuring of the imperial administration. It instituted obligatory conscription, modern education and, most importantly, proclaimed all Ottoman citizens equal in front of the Ottoman law. Importantly, the principle of equality abolished the previous system (*millet*), which had given full right to political participation of the Sunni Muslims, but subjected Christians and Jews to taxation and did not grant them full citizenship.

Various actors strived to affirm their own vision of what kind of restoration of order the *Tanzimāt* should bring. The Istanbul authorities envisioned

**Figure 1** Map of the present seat of war in Turkey and Syria. 1840. British Library, Maps 46970.(6.). The map shows the three possible lines in the region of Syria behind which Ibrahim Pasha was to withdraw as part of the Convention of London to end the Turco-Egyptian war and the invasion of Syria.

the Tanzimāt as a secular principle of Ottoman citizenship. Differently, however, the European powers read the Tanzimāt as a restoration of order in a sectarian sense (Makdisi 2000). Within these different interpretations of what order and modernity meant in the Ottoman empire, and within the social unsettlement that the Tanzimāt produced, the sect gradually established itself as the new unit of representation in the political landscape of Ottoman Mount Lebanon. Military and commercial cartographers now relied on population counts to draw new maps of Mount Lebanon taking into account, for the first time, religious belonging. One of the first such maps has been drawn by James Wyld in 1840.[7] This map contains a mix of commercial information, including cultivations, as well as a demographic table with a population count, by sectarian belonging.

Cartography goes hand in hand with the process of establishment of the nation state (Harley 1998). In the nineteenth century, the European powers were substantial patrons of map-making in this area of the Levant.

These maps, however, were not only technical tools to facilitate commerce and military or 'fact finding' expeditions: they also came with a baggage of assumptions about the sociality of a certain territory. In the case of the Ottoman Mount Lebanon, 'the silent lines of the paper landscape foster the notion of socially empty space' (Harley 1998, 134), over-imposing a sectarianized space over the genealogical geographies (Makdisi 2000) of the pre-Tanzimat era.

Within the context of the Tanzimāt, the traditional social relations of the provincial notables (*muqatā'ji*), with the commoners (*ahāli*) inhabiting their lands and with the Ottoman authority began to change. Before the Tanzimāt, the feudal lords exercised power of taxation and maintained the order autonomously from the direct power of the Sultan. Within this sphere of autonomy, the mountain chieftains had extended their power based on lineage, social prestige and wealth. The spatiality of pre-Tanzimāt Mount Lebanon can be described using Makdisi's definition of 'genealogical geography' (Makdisi 2000, 31) where specific power and legitimacy was 'more a function of personal loyalty between protector and protégé than an attribute of coercion or impersonal authority' (Khalaf 2002a, 65). Also the relations between the ruling families were structured around personal and inter-religious ties; this reflected spatially in the fact that the borders between the chieftains' lands were more like negotiable buffer zones rather than clear-cut territorial demarcations (Makdisi 2000). Once the Tanzimāt dissolved the spatially loose mechanisms for social distinction and privilege, however, the sect became not only the domain within which new discourses of legitimacy and power were constructed, but also the terrain on which new rivalries and disputes developed (Makdisi 2000).

In 1841, the first sectarian clashes broke out in the town of Deir al Qamar.[8] The novelty in this violence was about the prevalently sectarian character of the strife, regardless of cross-community loyalty that prevailed until recently: non-sectarian social geographies where 'a quietist religion and passive subjecthood [...]separate from a non-sectarian notable politics' (Makdisi 2000, 64) were replaced by 'the religious identity of the local inhabitants as he point of departure for a modern reformed and ambivalent Ottoman sovereignty in Mount Lebanon.[...]. In this new development the social, the political and the religious were explicitly [...] and antagonistically fused together' (Makdisi 2000, 64). In other words, in 1841 Druze nobles had attacked their own servants, severing previous relations of loyalty and proximity, and instead called upon other Druzes from other villages, in the name of the sect. To stop the clashes, the Ottoman authorities closely consulted with European diplomats (Farah 2000; Hakim 2013). In May 1842, the Ottoman ambassador to London Âli Efendi and the Austrian chancellor Metternich concluded that

> 'only three choices existed for governing the Lebanon. These were either to restore Bashir's rule, to place the Maronites under a price of their own and the Druzes under a Druze, or to place the Mountain under the direct rule of an Ottoman paşa. Âli and Metternich agreed that the second alternative would be the best, with "representatives of the sultan" arbitrating between Maronite and Druze governors when disputes arose' (Farah 2000, 187).

Spatially, this translated into the partition of Mount Lebanon into two Lieutenancies on the basis of sectarian belonging (Figure 2): a 'Druze Lebanon south of Dohr [sic] el Baidar and Christian Lebanon north of this [mountain] pass. To each section was assigned a distinct Caimacam [sic], druse for the supposed druse part, Christian for the other' (Charon 1905, 340). The solution, in other words, was a territorial separation of the religious communities from each other by creating spaces where sects could live 'orderly' side by side.

Accepting and approving the partition proposal entailed a great deal of diplomatic controversy and resistance from notables who were keen on preserving their privileges and lands. Most importantly, it was the territorial realization of the partition that revealed to be the greatest challenge. The two possible options were either a geographical division (with each Kaimakam administering both Christian and Druze in his own area), or a sectarian one (with each Kaimakam administering over all their coreligionist wherever they are (Command of Her Majesty 1845, 5; Farah 2000, chap. 10). While the first proposal would have implied problematic population transfers, the second was practically difficult to implement, especially in the most religiously mixed sectors (*muqata'at*).

A decision took years, and meanwhile communitarian tension in Mount Lebanon increased, revealing the complexity of the partition attempting to produce a coherent map out of what was instead a religiously mixed territory resulting from a pre-existing trans-communal social fabric. The partition was problematic because the sectarian diversity of the population of the ground did not fit the neat territories that the (mainly European) cartographers were representing as facts on the ground:

> 'the Lebanese population was not neatly divided geographically between a Christian and a Druze sector. In the Christian sector [...] only the northern districts and part of the central districts were inhabited solely by Christians. In the Matn, attached to the Christian sector, lived a small Druze minority, while in the southern Druze sector the Christians formed a slight majority' (Hakim 2013, 49).

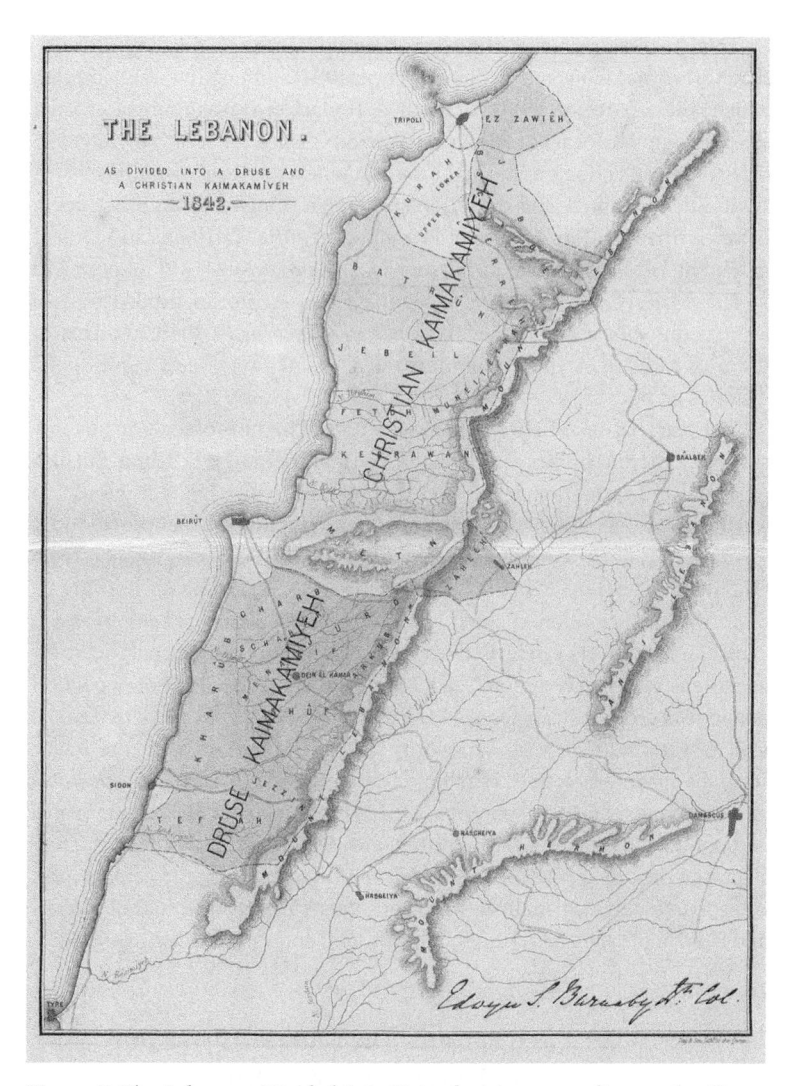

**Figure 2** The Lebanon. Divided into Kaimakamiyes according to the Commissioner's 1st project of XLVII Articles. 1861. The National Archives, Kew. FO 925/2889. This map was produced as part of the works of the commission of European powers tasked with investigating the 1860 massacres in Mount Lebanon and Damascus. It shows the status quo in 1842, when Mount Lebanon was split into two Lieutenancies, later ratified with the Reglement (*düzeni*) of Şekib Effendi.

This epoch was punctuated by continuous adjustments to the administrative structure of the Lieutenancies (Tarazi-Fawaz 1994) and lengthy discussion as to which kind of partition should be reached. The lack of a resolution exacerbated the occurrence of local skirmishes in its mixed villages. It was only in May 1845 that an official ruling on the dual Kaimakamate - the Reglement (*düzeni*) of Şekib Effendi named after the Ottoman Foreign Minister – was proclaimed. It was written in 'Ten Articles of Peace between the Christians and Druses of Mount Lebanon, signed at Beyrout [sic] the 31st May 1845' (Command of Her Majesty 1845, 5). This recognized two territories headed by two Kaimakams, but also assigned Wakils (agents) to assist the Christians of the mixed districts, and an administrative council to resolve confessional disputes on both sides.

The years between 1840 and 1845 were crucial in the proliferation not only of diplomatic and military intervention to stabilize Mount Lebanon, but also of spatial and cartographic practices that worked to intertwine religious and territorial belonging. The Reglement of Shakib Effendi 'formally introduced for the first time the communal factor at the political and institutional levels' (Hakim 2013, 52).

In this configuration, the notion of 'coexistence' became a key concept to explain relations between sects: the double imaginary of distinct, but pacifically coexistent, religious communities of Mount Lebanon acquired now a political and indeed geographical presence. The partition reflected the modern colonial idea of European 'rational superiority' above the ambiguity of the multi-religious spaces of Mount Lebanon. Through the European power apparatus and the Western-inspired governance of the Ottoman authorities, the spatial and political constitution of Mount Lebanon acquired sectarian character also thanks to the cartographies of enclosure, boundary-tracing, and demographic categorization, which contributed to consolidate new geographies of identity and alterity, of safety and danger, of loyalty and enmity (Makdisi 2000).

## The sect as dispositif. The Mutasarrifiyya of Mount Lebanon (1860–1920)

If the geographical partition of Mount Lebanon did not envisage a hard border, the distinction between the two Lieutenancies was soon to be marked by a tangible infrastructural boundary. In 1857, the old Beirut-Damascus road, until then 'a path served, above all, by mules and camels' (Eleftériadès 1944, 37), was upgraded to a large carriageway for fast diligences. But reorganizing the territory and population of lands until then structured according to flexible multireligious relations was no easy task, and it had hard consequences less than twenty years later.

Incidents and simmering communal tension throughout the 1840s and 1850s escalated into major violence in and beyond Mount Lebanon in 1860, through the Bekaa Valley, and reaching Damascus. The prelude to the violence was a revolt by Maronite commoners in the Keserwan region of the Maronite Kaymakamate who – often inspired by the clergy – rose against the authority of their overlords. Here, pacific gatherings turned into radicalized and armed confrontations, until violent revolts against the feudal lords broke out. These events subverted the traditional relationships based on the loyalty of the commoners to their lords. For the first time, the commoners entered the realm of politics, and in so doing they used the Maronite sect as a frame for their newly found political identification and action:

> The extreme and historically unique violence in 1860 in Mount Lebanon is intrinsically connected to the transformation of political subjectivity specifically the transformation of the ahali from passive, ignorant, and loyal 'imperial subjects' from objects of the Sultan and lord, to active, political subjects. (Kastrinou 2016, 10)

Social revolt was strongly formulated in sectarian terms and against the Druzes as it extended to the southern districts of the Maronite Kaymakamate, in the Metn district, with skirmishes and cycles of revenge between Maronite and Druze residents causing heavy loss of life. Gradually, what is irregular and often opportunistic violence acquired a systematic, planned and wanton nature and became a fully-fledged civil war. Makdisi (2000)[9] reports incidents between 1858 and 1860 involving the built fabric of villages, including cutting down trees, looting, destroying harvests and houses, razing of farms, and the mutual burning of sacred sites and monasteries. The feudal socio-spatial relations of genealogy and proximity that had constituted the old emirate order were being inexorably overcome. Sect was now what animated new political subjectivities, and in the name of the sect, the spatial expression of that old order was being materially and inexorably undone. New geographies were being traced, where 'security was guaranteed by the absence of rival communities' (Makdisi 2000, 134). A new view of a national territory, based on religion, would have not made sense in the pre-Tanzimāt era. Now, instead, reinforced by the twenty-year experience of the Double Kaimakamate, the sect had become a new episteme and a new geography: a new way of envisaging the political self, and of organizing territory. Doing away with the traditional social privileges and relations that constituted the pre-Tanzimat ancient regime, commoners took matters into their own hands, acting in the name of their sect to protect co-religionaries where diplomacy was failing (Makdisi 2000, chap. 7):

> they imagined a Christian geography that bound villagers to Kisrawan to those of the Matn and the Shouf; in so doing they defied the traditional demarcations of the notable families that had hitherto defined Lebanese geography, and they ignored the new geography of the European-Ottoman partition that delegated authority to kaymakams chosen from among the elites. (Makdisi 2000: 120)

Much has been written on the 1860 violence (Farah 2000; Tarazi-Fawaz 1988, 1994; Hakim 2013; Kastrinou 2016, chap. Introduction; U. Makdisi 2000; Rodogno 2012; Shehadi and Mills 1988) and it is not the aim of this chapter to offer a chronicle of the events. What is important to underline is that the massacres that took place across villages in Mount Lebanon used the sect as a dispositif for modern and national political identification, and that in turn underpinned the violent purification of specific territories and built environments from the members of the opposing sect. The wanton destruction of houses and crops in the towns of Hasbaya, Rashaya, Zahleh, and Deir al Amar are just few examples of how the built environment of the rural mountain in the 1860s became a target of identitarian violence aimed at the erasure of the physical space of the rival sect. At the sight of the Druze-majority town of Rashaya in the aftermath of the 1860 conflict, the then Prince of Wales wrote that 'In this town, 400 to 500 Christians were massacred and we saw still the remains of the burnt houses' (Figure 3).

Under the pressure of an indignant Western public opinion, an international commission[10] was created on 5 October 1860 (Hakim 2013) to investigate the causes of the massacres, revise the Reglement of Chekib Effendi that had instituted the problematic territorial entity of the Double Kaimakamate and propose ways to avoid future violence. Once again, the territory of Mount Lebanon had to be redesigned in order to pacify the rivalries brought about by the new political identification in the sect. The commission worked on several different spatial visions for the province, discussed at meetings and circulated among the population who often responsed (Hakim (2013). The work of the five commissioners was not without internal controversy. Particularly, a divide arose, in the first phases of the works of the commission, between the French and English commissioners, which transpired in the two main projects for the reorganization of the area that were presented to the European and Ottoman authorities. Each was criticized and opposed, respectively, by the French and by the British commissioners. The first proposal, by the British commissioner Lord Dufferin, and objected by the French commissioner Leon Béclard,

**Figure 3** Rasheya [Rashaya, Lebanon], 27 April 1862. By Francis Bedford (1815–1894). Royal Collection Trust/© Her Majesty Queen Elizabeth II 2018. Catalogue No: RCIN 2700955.

stemmed from the territorial layout of the Double Kaimakamate. 'provided for the establishment of three separate districts, one for each of the maronite, Druze and Greek Orthodox communities, and the segregation of the populations of the Mountain in order to fit these new administrative units' (Hakim 2013, 73). The map of the proposal contains the details of how the segregation had to be implemented. This includes population exchanges between the Druze population of the Metn and the Christian population of the mixed districts, as well as property transfer in the areas with Druze property but a majority Christian population; special administration for Zalhe and Deir al Qamar; and the annexation into Mount Lebanon of the Lower el Kurah district as a Greek Orthodox Kaimakamate. The second proposal, devised by the French commissioner and objected by Dufferin, was instead based on the pre-1840 territorial situation and envisaged a united Mount Lebanon under a Christian Governor, using the map designed by the French Expeditionary Force, led by General Beaufort d' Hautpoul that had landed in Beirut in August of the same year. As Hakim (2013) notes, 'the main interest of this plan is that it marked the first appearance on a map of the core of what was to become Greater Lebanon some sixty years later, along with a whole historical legitimising view' (86).

Despite the cartographic controversy, the common question running through the diverse map proposals, was: which territorial formula would best reflect the changing political subjectivities in Mount Lebanon in a peaceful way? The political change put in motion with the fall of the Shehab Emirate, the Tanzimat, as well as the envisioning of the population and its space in sectarian terms with the Kaimakamate – all under the influence of Europe's economic and cultural colonial power-now needed a territory on which to function for the long term. The question of sectarianism, therefore, is at the very basis of the formation of Lebanon as a nation state as a socio-cultural entity, but also, and most importantly, as a territory with a government and borders. It is between 1840 and 1860 that the sect becomes an essential premise for any territorial project in this area of the Levant. The question of which territorial shape a sovereign state of Lebanon should take, as we will see, will return with extreme force in the first phases of the civil war and during the partition of Beirut. The motor of that partition is not the presumably inescapable nature of sectarian belonging in Lebanon, but rather, the construction and normalization of sectarianism as part of a modern territorial project initiated in post-Shehab Mount Lebanon. A project designed to reorganise social relations amidst tectonic changes in the ways of seeing politics and identity that took place in the region as part of colonial modernity.

The commissioners proposed several plans, but a resolution to the reorganization of territory was only achieved in March 1861, not without remarkable differences from the initial plans. Dufferin had come to believe that union, rather than partition, was the way to avoid future violence. He now proposed strong and centralized government for Syria, independent from Istanbul, where sectarian divisions – now deemed as 'barbarous distinctions which have hitherto divided its inhabitants into innumerable tribes and sects' – 'ought not be converted into a geographical expression' (Dufferin, quoted in Hakim 2013, 76). This was now opposed by the French commissioner, who wanted instead to preserve the Christian privileges of the area, by separating it from Syria and appointing a Christian governor. Dufferin responded with conceding that Christian-majority northern Mount Lebanon preserves its autonomy under the Kaimakam, 'while leaving the rest of the Mountain under the same government as the remaining Syria provinces' (Hakim 2013, 77). The plan that was eventually approved was an autonomous mountain region, with annexed the mixed districts of the southern Mount Lebanon, under by a non-native Christian governor, equal to any other Pasha of the Ottoman empire.

The commission's initial ambition of a reorganization of Syria had faded by the end of the negotiations, at least in part. While public opinion in Europe was adamant to bring profound changes to the area to avoid a recurring of violence, the officials in the commission were more cautiously looking for compromise, as they started to appreciate the complexity of the situation and they wished to avoid controversies that would re-open the Eastern Question, especially in light of the recent Crimean War. Nevertheless, this was a crucial moment in the formation of the Lebanese nation: 'all the nationalist ideas and options that would be adopted by some local party or other, from 1860 until 1920, were formulated and examined at this time' (Hakim 2013, 74). Not only the territorial shape of Lebanon started to be determined, but, and even more importantly, this is when we see the establishment of a specific relationship between sect, government, and territorial organization in the region – a sectarian *dispositif*.

A new territory, the *Mutasarifiyya* (Governatorate) of Mount Lebanon, was instituted upon decision of the Commission and the Ottoman authorities, abolishing the previous system of the Double Kaimakamate. In 1861, the Ottoman authorities issued an administrative document called *Réglement Organique*[11] sanctioning the new regime. The status of Mutasarrifiyya was not unique to Mount Lebanon in the Ottoman Levant,[12] but differently from the other provinces of the Ottoman empire, it had greater authority from the Sultan of Istanbul who nominated a non-Lebanese Christian to the role of governor (*Mutasarrif*). The Mutasarrif was appointed by and responded to the Porte, had executive, administrative and judicial powers and was assisted by an Administrative Council – elected by the village sheikhs – representing the six main religious communities of Mount Lebanon. The Ottoman police was now the only institution entitled to use violence to guarantee the security of all the citizens, regardless of their previous notable affiliations. The territory was divided into seven Qadas (districtis), with standardized taxation and each under a Kaimakam appointed by the governor from the dominant sect in the Qada.

Governing the Mutasarrifiya meant doing away with a problematic, and perhaps impracticable, geographical partition of the sects, but it also meant embedding the sect more subtly and more pervasively, as a dispositif of security and sovereignty within a unified territory. Sectarian belonging in the Mutasarifiyya determined the electoral representation in the six provinces as well as the access of the individuals to public functions. The genealogical geographies that constituted the social fabric of Mount Lebanon until thirty years earlier (Makdisi 2000) were now officially replaced by regulations that 'enshrined for the next century and a half (up until the present time) the principle of government by sectarian appointment' (Farah 2000, xx–xxi).

## Normalizing the sect: The French mandate and the Lebanese Republic (1920–1946)

The regime of the Mutasarrifiyya ended after the First World War, together with the Ottoman Empire. At the 1920 San Remo Conference, the Allied governments of Britain, France and Italy re-drew the political boundaries of the Ottoman lands after the defeat of the Empire and the Central Powers. New nation states were born while the Ottoman Empire's territory and social composition were undergoing tectonic changes. On one hand, the Treaty of Sevres split the core of the Ottoman empire (Anatolia) and opened a confrontation for the determination of Turkish sovereignty between the Ottoman government and the Kemalists that resulted in extraordinary violence against Kurds, Armenians and Greeks of Anatolia and the disintegration of the Sultanate in 1922 (Rogan 2016). On the other hand, the French authorities proclaimed the Republic of Greater Lebanon (*Grand Liban*) under its colonial mandate on the basis of the Sykes-Picot agreement between the French and British governments for the partition of the Middle East.[13] After the battle of Maysalun against King Faysal of Jordan, General Gouraud annexed the Bekaa Valley to the Vilayet of Beirut and the former Mutasarrifiyya of Mount Lebanon and proclaimed the state of Greater Lebanon on 1 September 1920. In the same way in which the Maronite and the Druze were divided earlier in the mountains, the French government tried to break the relations between Shiite and Sunni Muslims in the southern provinces of the new state and to gain the support of the Shiite, against the Sunni mainly pan-Arab political preferences. This caused a divergence of territorial visions around the same political reality among the Shiite masses, notables, and clergy, many of whom continued supporting a vision of Lebanon as part of a greater Arab nation (Firro 2006), a division which would persist and return in the urban geopolitics of the 1975–1976 civil war.

The creation of Greater Lebanon was a lengthy process and full of setbacks and contrasting positions: while on the surface of the debate the main separation was between the (Britain-backed and mainly Muslim) advocates of an independent and centralized Syria, and a mostly Christian and France-backed vision of a separate Greater Lebanon, what in fact underpinned this neat division was 'a wild diversity of social an communal cleavages within each camp, encompassing a vast array of often conflicting local interests and loyalties' (Hakim 2013, 230). But ultimately, the nationalist ideas and their geographical translations that developed at the time of the works of the European Commission into the massacres of 1860s, and especially reports by activists like Jouplain (Hakim 2013) advocating the expansion of the territory of Mount Lebanon much beyond the Mutasarrifiyya, became now more relevant than ever before. As mentioned, the cartographic evidence to support the argument for the establishment of Greater Lebanon with an enlarged territory already existed as '[Beaufort's] project for a Greater Lebanon, and the map drawn under his auspices, disregarded at that time, resurfaced in 1919 when it was adopted by the Lebanese nationalist and the Maronite Patriarch as a yardstick for the frontiers of the Greater Lebanon they were claiming' (Hakim 2013, 87). The sect was now the normalized part and parcel of the French mandate government in Greater Lebanon, and statistical knowledge about religion, much like in the maps drawn by the European powers to bring cartographic order out of the 'saraband of races'[14] in the Ottoman Levant, was paramount to the functioning and governing of the new nation state. In much the same way, the mandate government used the imaginative geographies of a 'fluctuating' population and of disorder in this part of the Mediterranean Levant to shape a rational space for modern Lebanon that served to represent a national territory through the enumeration, visualization, and exhibition of its features, among which were the religious communities:

> The extent of the territories, the primitive state of the native tribes, illiteracy, the fact that in some cases the population fluctuates (since it comprises nomadic or semi-nomadic tribes), and migratory movements are all obstacles which render the task of the Administration in this respect peculiarly difficult. […] in most of the territories in question, all attempts to count the population, to establish a register of births, marriages, and deaths, to register the inhabitants or prepare accurate population statistics, or even approximate estimates, encounter many difficulties. (League of Nations 1945)

In 1926, a republican constitution, strongly inspired by the French and Belgian ones, was adopted. The parliament was based on a sectarian division of seats, although in the administrative districts the election would not follow a sectarian pattern: a candidate would obtain the majority of the votes in their district, regardless of the sect which the electors belong to. Nevertheless, this practice lent itself to gerrymandering, in order to advantage specific candidates by designing homogeneous districts from a religious point of view (Corm 2005). Since 1926, the mandate authorities and the independent Lebanese government engaged in a process which gradually regulated the legal life of the sects, granting that issues of personal and family law, as well as education, could be dealt with by religious tribunals. In 1936, France crystallized the role of the religious community as the main unit of political organization through two organic laws (*arrêtés*) issued by the French High Commissioner. The documents stated the existence of seventeen 'historical [sectarian] communities' (*communautés historiques*) in Lebanon.[15] Through the idea of *communauté historique*, religion was now a dispositif for the circulation and mobilization of specific political identification and action, as it had importantly become between the 1840s and 1860s. Most importantly, it was also a source of law and the unit of legitimacy for the highest political powers.

## Overlapping territories. The mountain and the shifting urban geopolitics of Beirut

Tectonic shifts in social privilege, land ownership and taxation underpinned socio-political tensions that escalated into civil war in Mount Lebanon since the end of the Egyptian campaign and the fall of the Shihab Emirate in the late 1830s. In this landscape, the sect gradually became the cement for the formation of new political identities that came to clash violently. Importantly, however, tensions also simmered in a climate of economic decline and indebtment, especially among local silk manufacturers (Tarazi-Fawaz 1988, 1994). New, Europe-backed policies in silk production turned Mount Lebanon predominantly into an exporter of raw material, rather than a manufacturer for the Ottoman market as it had previously been, thus becoming economically dependent on European markets as buyers via a constellation of dragomans, merchants and middlemen based in Beirut (Tarazi-Fawaz 1988), all amidst a 'massive shift of trade from the hinterland to the coastal areas' (Tarazi-Fawaz 1988, 65).

These developments in the Mountain were by no means limited to their immediate locales. In 1860, sectarian conflict quickly spread from the Metn to Hasbaya and Rashaya, crossing the Beqaa to Damascus, and the geopolitical ramifications at the very root of that violence – the power game between the Ottoman empire and the European powers at least since the Egyptian campaign in Syria – reached far and wide beyond the confines of the mountain. The situation in Mount Lebanon provided the context for the changing urban geopolitics of Beirut on the coast, in ways that would at least in part help to historically contextualize the emergence of specific urban fault lines during the civil war. Beirut had already started to attract foreign consuls and was home to a 'merchant aristocracy' (Saliba and Assaf 1998, 7) around its port since the 1830s as the capital of the Vilayet of Sidon. However, it is only between the 1840s and 1860s that the city underwent a major growth in population and activity (Boudjikanian-Keuroghlian 1994; Tarazi-Fawaz 1988), to eventually become the capital of the nation state of Greater Lebanon under French mandate in 1920. Saliba and Assaf (1998, 11) report that Beirut's size increased '13 times, from around 150,000 m2 in 1940 to 2,000,000 m2 in 1875'. Beirut's population grew from 7000 inhabitants in 1820, to 10000 in 1840, to then quadruplicate in 1860, when it reached 40000. Between 1860 and 1880, the city's population further doubled to 80000, and reached 130,000 between 1880 and 1920 (Saliba and Assaf 1998).[16] At a time when European commissioners, military expeditionary corps, and diplomats were mapping the Mountain, counting its population by religious belonging, partitioning it and reorganizing it geographically to resolve a decades-long

long spiralling of conflict, the continuing violence and economic decline caused a substantial wave of migration from Mount Lebanon and beyond, including Damascus and Aleppo (Tarazi-Fawaz 1994) into Beirut which, especially after the stationing of the French Expedition in 1860, had become a more secure urban environment (Davie 1992). The city thus changed dramatically in size, population and density and projected itself towards becoming an economic and political capital: 'It was between 1840 and 1860 that Beirut became an important commercial, financial and diplomatic centre. Moreover, it linked for the first time the fortunes of Mount Lebanon, as a silk producer, to its own cycle of economic and political activities' (Tarazi-Fawaz 1988, 65). In particular, the decline of silk production and trade which, until the 1850s, flourished in Mount Lebanon, started a vast wave of mainly Christian (Boudjikanian-Keuroghlian 1994; Ruppert and Verdeil 1999) migration towards the coastal cities. As Fawaz notes, 'it is this new role of Beirut which eroded the isolation of the Mountain as an enclave of heretic and religious minorities' (Tarazi-Fawaz 1988, 65). We can now better contextualize Davie's (1992) claims that topographic and transport considerations, together with a preference for proximity to co-religionaries, and to providers of socio-economic welfare and education like universities and foreign missions, determined a certain communal homogeneity in the settlement patterns of the refugees from the 1840–1860 violence in the city. These patterns became, to some degree, spatial constraints preventing the tangible realization of the equal 'ottoman citizen' of the Tanzimat (Davie 1992) and reinforced instead the confessional character of specific neighbourhoods in Beirut.

In 1888, the creation of the Vilayet of Beirut (containing the Sanjaks of Latakia, Tripoli, Beirut, Akka, and Nablus) was a recognition of its growing importance and strategic status by the Ottoman empire, as well as an attempt 'to reinforce [the latter's] control, in order to fight against the growing European influence and the parallel emergence of local identities' (Eddé 2009, 20). Beirut continued to grow at least until the First World War, when a reverse migration, especially Maronite, from Beirut back to the Mutasarrifiyya of Mount Lebanon, which was exempt from conscription, caused the shrinking of Beirut's population, due to migration, famine and bombardments by foreign navies. In the 1920s, further waves of displaced people, such as Armenians from Turkey and Greeks from Anatolia, fled to the Lebanese coast and especially to Beirut in order to escape the ethnic and religious massacres taking place in other areas of the Ottoman Empire. These factors, together with the arrival of Palestinian refugees from the newly declared state of Israel in 1948, were the shock waves which modified the urban landscape and political culture of Lebanon's new urban centre.

Distinct territorial visions underpinned the discursive and representational struggle for the shaping of the Lebanese nation. The main divide lay between a political and cultural imaginary that saw Greater Lebanon as integral part of the Arab nation, and one that instead viewed Lebanon as an individual state with clear borders, in light of its (Christian) cultural specificity. The first imaginary was mainly (but not totally) supported by the Muslim (and particularly Sunni) population, while the Christians – especially the new Maronite bourgeoisie who had emigrated from the Mountain and flourished in Beirut during the mandate – were advocates of the Lebanist thesis, which looked with favourable eyes to the West and especially to France (Firro 2003). The Maronites were the ones whose sense of religious community overlapped and even coincided with their national vision (el Khazen 2000): they conceived the new state of Greater Lebanon as the result of their struggles dating back to the 1860s with the institution of the Mutasarrifiyya following the Maronite insurgency in Mount Lebanon (Corm 2005). Diverging territorial understandings of the Lebanese state, inspired by different nationalist ideas that had emerged during the mandate (Firro 2003), developed and occasionally clashed, especially in Beirut, and 'from the 1930s, the streets of Beirut, every now and then, became the scene of violent clashes between Christian and Muslim gangs, one side brandishing the banner of Lebanism, the other of Arabism' (Salibi 1988, 180).

The sorts of Beirut and Mount Lebanon are spatially, politically and geopolitically intertwined. The economic and political changes in the Lebanese mountain between 1840 and 1860 had deep consequences for the morphology and social fabric of Beirut. From the partition of Mount Lebanon into two lieutenancies, to the proposals for the rearrangement of the region by the European Commission and the controversies over the spatial extent and meaning of Greater Lebanon, the modern Lebanese nation state and its sovereignty have been built around a spatial and political project of coexistence within religious difference. The Taifa has been part and parcel of a process of determination of and agreement over the territorial shape of Lebanon's sovereignty that, as will later see, was far from settled at the start of the civil war in 1975.

The story of the conflict in Beirut in 1975 and 1976 starts therefore much earlier, in Mount Lebanon. The question of sovereignty and that of Taifa have overlapped since at least 1840 and any serious discussion about urban conflict in Beirut cannot avoid considering these overlaps. The political violence in the mountain was far from being an exclusively local affair: its echoes reached Europe and beyond, and what followed – the European Commission investigating the clashes – was one of the first international humanitarian, military, diplomatic and cartographic interventions in what were the first of a series of sectarian and ethnic waves of violence in a rapidly changing Eastern Mediterranean at the dusk of the Ottoman empire around the First World War. Beirut's functional and political growth at the turn of the century was directly connected to the events in the Mountain: the city hosted economic migrants, housed merchants and dragomans, settled refugees form the mountains, and provided the space for the diplomatic missions, military expeditions, and all the official personnel involved in dealing with the Mount Lebanon crisis. Any subsequent urban conflict in Beirut, down to individual buildings, must be seen in these wider geopolitical, trans-scalar and postcolonial connections. The next chapter contextualizes how local and international understandings of Lebanon's sovereignty have developed in the post-independence phase, and prompted external military intervention to curb intra-state conflict. Particularly, it observes how Cold War regional geopolitics played out on the urban ground, and how specific urban geographies of violence contributed to shape the ways sovereignty was represented and, most importantly, acted upon militarily.

# 3

# Lebanon Salvaged: Sovereignty and Urban Space in the Republic of Lebanon (1943–1975)

After its suspension under the Vichy regime during the Second World War, and following popular uprising, the Lebanese constitution was re-instated and independence was proclaimed by the commander in chief of the Free French forces in the Middle East, General George Catroux on 22 November 1943. This came after a fortnight of street protests, the imprisonment of president of the Republic Bishara al Khoury and Prime Minister Riad al Solh, and the secretive but crucial alliance of the latter with General Edward Spears, British minister for Syria and Lebanon and head of the British mission in Syria (Zamir 2005). This was a phase in which different nationalist visions spoke in the name of a 'rational' dialogue about national belonging, although their interpretation of the idea of the Lebanese nation and its sovereignty was highly contested (Salibi 2003).

For example, the National Bloc, a Christian political grouping headed by Egyptian-born Raymond Eddé, demanded that a special relationship with France be maintained after the end of the Mandate, for fear that – if not properly handled – independence might imply a risk for territorial integrity. The more religiously heterogeneous Constitutional Bloc instead was enthusiast about full-scale independence. The Constitutional Bloc, gathering mainly urban Christians, Shiite and Druze, awaited independence with anticipation because of the economic dynamism it could bring. Other groups demanded that 'Lebanon be made a national home for the Christians under French protection, just as Palestine was to be made a national home for the Jews' (Salibi 2003, 184) and proposed to integrate the Muslims into Syrian territory. Pan-Arab nationalists – spread mainly among the Sunni – also relished independence for this reason.

Eventually, a pragmatic compromise was reached in 1943 when the National Pact (*al mithāq al watani*) was stipulated between the president and the prime minister. Through the pact, the Maronite and Sunni communities had reached a compromise on power sharing: the Maronite renounced French protection as a legacy of the mandate and recognized the Arab cultural roots of

Lebanon. The Sunni, on their part, renounced their claims to integrate Lebanon territorially in the Syrian hinterland and accepted instead the borders traced by France in 1920 when the colonial mandate was instituted. The National Pact also distributed the highest institutional powers among the main religious groups: the presidency of the Republic was to go to a Christian Maronite, the presidency of the cabinet to a Sunni Muslim, and that of the parliamentary chambers to a Shia Muslim. This division was based on a population census held in 1932 – the last official census taken in the country until the present day – which showed a slight Christian majority. The importance of the National Pact resides in that the notion of coexistence and sectarian balance in Lebanese public life was now institutionalized. This pact is still considered as the seal on Lebanon's existence as a modern consociation democracy where religious communities, although different among themselves, can coexist in peace. Although many consider the pact as a shaky compromise based on a double negation of Maronite and Sunni claims (el Khazen 2000) which brought no constructive policy for the country, it is also important to remark on a fundamental aspect of the pact for Lebanon's representation as a modern nation state: the pact effectively sanctioned the political equivalence between sectarianism and the very existence of Lebanon as a modern territorial nation state. It officially seals the notion of sectarian coexistence, exactly one century after the notion of Taifa was first translated geographically into the double Kaimakamate and became the main platform for expression and mobilization of new, modern forms of political subjectivity and national identification, breaking away from the *ancien regime* of the Ottoman administration.

This political consecration of the notion of coexistence, much like the idea of haven and refuge during the clashes in Mount Lebanon less than one century earlier, translated into differing national and territorial visions for Lebanon. The idea of refuge was largely a production of the imperial imaginative geographies about Lebanon from the early nineteenth century and was also used by nationalist (and often Christian) leaders since the 1920s, seeking to promote nationalist theses about Lebanon as a land with the specificity of refuge for the Eastern Christians and, therefore, deserving to remain a separate entity from the rest of the Arab lands. The negation of that refuge imaginary came instead from the supporters of pan-Arabism, who envisioned Lebanon as integral part of its Arab surroundings. These territorial considerations became gradually more significant in 1958, when the United Arab Republic (UAR)[1] of Syria and Egypt was proclaimed. Now, the tension between the two ideas of Lebanon became prominent as Arabism acquired a concrete territorial expression in the UAR, which, in turn, reinforced the already existing Christian fear of Sunni Arabism (Salibi 1988) in Lebanon.

## Geopolitical scripts of fragility

Two crises marked the Lebanese political scene after the National Pact. The first was a brief moment of sectarian tension in 1952, followed by the resignation of President Bishara al Khoury and his replacement by National Liberal Party's leader Michel Chamoun.[2] The second one, longer and more violent, saw supporters of pan-Arabism and Lebanism clashing in Beirut, bringing the compromise of the National Pact to a critical trial. The crisis in 1958 marked the first foreign military intervention in post-independence Lebanon. In the summer of that year, President Camille Chamoun called for US assistance to curb civil strife in Beirut, invoking the Eisenhower Doctrine[3] in the name of protecting Lebanon's sovereignty. The crisis was contemporary to the fall of the Iraqi monarchy after a state coup by the socialist Ba'ath party and to the fusion of Syria and Egypt in the UAR. The 1958 events have been described by some as an internal affair – the clash of two different interpretations of Lebanese history (Salibi 1976) – and by others as the culmination of external geopolitical factors within the wider Cold War geopolitical scenario (Corm 2005; el-Khazen 2000). However, both versions tend to idealize and disembody violence and, more specifically, consider the localized implications of the crisis as a mere backdrop for the unfolding of political events.

Geopolitical discourse is never only strategic; it contains idealism, moral values used to construct place according to specific imaginations (Ó Tuathail 1992; Gregory 1995). The term 'script' here indicates 'a set of representations, a collection of descriptions, scenarios and attributes which are deemed relevant and appropriate to defining and judging place within reasonings of foreign policy' (Ó Tuathail 1992, 156). Much like a theatrical script, a geopolitical script is 'a regularized way of acting and talking when negotiating certain [geopolitical] situations' (Atkinson 2005, 69). When a geopolitical script becomes predominant, it is often accepted as a 'value-neutral description' (Sharp 1998, 159). In this view, events of apparently similar nature (e.g. intra-state conflict in 1958 and 1975–1976) acquire different understandings in foreign policy and attract different (political and military) responses, depending on their *scripting* at a specific point in time. As we will see, Lebanon's sovereignty has been scripted in very different ways across different crises at different geopolitical conjunctures. But the localized and more precisely urban implications of these scriptings have rarely been explored.

Sovereignty is traditionally defined as the state's exclusive claim to political authority within a bordered territory, and control of the means of violence (Giddens 1985). In Lebanon's case, the notions of fragility and weakness have shaped the understanding of and action upon Lebanon's sovereignty in international foreign policy across its mandate and independence years. Lebanon is a consociation democracy often been described as resting on a fragile balance of power between its religious communities. One of the most famous Lebanese political thinkers, Michel Chiha, described the compromise of power between the religious communities, which lay at the heart of the National Pact, as a 'necessary' flaw to keep the country's religious and political diversity under control. He wrote: it is 'better for [Lebanon] to live with a limp than to break his back' (Salam 2001, 73, translation by author). Similar metaphors of weakness were used to describe Lebanon's post-independence economic development in the 1950s and 1960s. With an economy based on the service sector and on a laissez-faire presence of the state in economic matters, Lebanon has been described as a 'soft state' (*état mou*) with a 'sponge-like population' (*peuple éponge*) exposed to the influences of foreign projects and ideals (Corm 2005, 24), a country with a 'low degree of "stateness"' (Farid El-Khazen 2000, 92).

It is not my aim to produce a linear narrative of the 1958 crisis, as comprehensive and detailed political studies already exist (Agwani 1965; Alin 1994; Nawaf Salam 1979; Stewart 1958). Instead, I want to focus on the mechanisms used by the United States to script Lebanon geopolitically in the build-up and the aftermath of the crisis. These mechanisms include the scripting of Lebanon as part of a region 'on the edge', an amicable state incapable of defending itself from its surroundings due to its small territory, and Lebanon as a potentially dangerous quagmire, characterized by an ambiguity that the urban geopolitics of fighting in Beirut contributed to shape in US diplomatic environments. This chapter contextualizes the geopolitical scripts that underpin the views of Lebanon's sovereignty in the years preceding the civil war. This is important not only to understand how foreign policy viewed Lebanon's sovereignty and how this understanding then influenced its response to intra-state conflict, but it also highlights how specific imaginative geographies (Gregory 1995; Gregory 2004; Said 1995) about Beirut's urban space contributed to shape specific foreign policy understandings, performances and practices.

Two main understandings of Lebanese sovereignty emerged in official US foreign policy and beyond in 1958. Until the end of the Second World War, the United States respected the political and military supremacy of superpowers like France and Great Britain in the colonial Middle East. In Lebanon, religious, educational, and philanthropic institutions, such as the Syrian Protestant College (which then became the American University of Beirut), characterized the US presence. With the beginning of the Cold War and the decline of the European superpowers' presence in the region however, the US political weight in the Middle East increased, partly to fill a potential power vacuum caused by the decline of Great Britain's and France's roles in the region. The 1957 Eisenhower Doctrine itself was designed to counteract pan-Arab anti-Western hostility in the Arab world, and especially Soviet influence over Egypt and Syria in the wake of the Suez crisis.

The scripting of Lebanon as a place worth of American military intervention needs to be contextualized within the wider geopolitics of the Eastern Mediterranean in which, besides the official contexts of the Eisenhower Doctrine, 'oil and politics were key factors' (Gendzier 2006, xiv) to influence the US decision to intervene. Not only pragmatic commercial interests were at play in the US geopolitical script over the Middle East and Lebanon in particular, but these interests depended on the development of crucial physical infrastructures in and around Beirut. Keeping Lebanon politically stable at this point in time was paramount for the oil exporting economies of the Gulf, as well as for the United States. What is often considered as Beirut's 'Golden age', as the city turned into a regional economic and service hub and a haven of cosmopolitanism amid an authoritarian Arab hinterland, is actually underpinned by specific geopolitical economies and infrastructures. As a crucial area for US economic interests, it was unsurprising that Lebanon and the Mediterranean Levant became the pivot for two defining documents of US foreign intervention: the Truman Doctrine[4] in 1946 and the Eisenhower Doctrine. American intervention in the Eastern Mediterranean had already been requested by the British in 1946 when Greece and Turkey both received US economic aid under the Truman Doctrine to counteract Tito's and Stalin's interests in the two countries. At the same time, anti-British sentiment loomed large among British Greek-speaking citizens in Cyprus where, in 1958, the polarization between Greek and Turkish citizens exploded into communal violence, sending shock waves across mainland Greece. In the same year and alongside the nationalization of the Suez Canal, the UAR contributed to the Western fears of growing pro-Soviet Arab nationalism translating into territorial change. Regional tensions culminated in 1958 with the toppling of the Iraqi monarchy by the Ba'ath party. The script of Lebanon as a place whose sovereignty needs to be salvaged needs to be viewed in this wider Mediterranean context of containment of pan-Arabism and communism in order to protect US interests. The changing urban fabric of Beirut was a grounded everyday manifestation of these geopolitical processes.

In July that year, President Camille Chamoun (elected after the 1952 crisis) called for US assistance to curb protracted civil strife in Beirut under the Eisenhower Doctrine, which would be now implemented for the first time through Operation Blue Bat.[5] US President Dwight Eisenhower's administration sent 1500 marines from the US Sixth Fleet in the Mediterranean into inner Beirut to curb civil strife that had been protracting for weeks. The official scope of the mission was to

> protect American lives and by their presence to assist the Government of Lebanon in the preservation of Lebanon's territorial integrity and independence, which have been deemed vital to United States national interests and world peace. (Eisenhower 1958b)

Lebanon's 1958 strife can be seen as a conjunction of internal communal problems and 'the country's growing sensitivity to external shocks' (Kassir 2010: 453). On the one hand, deep social inequalities and divisions, as Kassir put it, 'became all the more acute as the regional conflict between Arab nationalism and Western foreign policy now began to infiltrate local politics' (453). On the other hand, the Cold War strategy of containment in the 1950s, as seen above, heavily involved Mediterranean states with high stakes for US oil-importing economy, that the United States considered 'on the edge' of falling into the Soviet sphere of influence or of Arab nationalism advocated in non-aligned Egypt under Nasser's guide. As a consequence, Lebanon was geopolitically scripted as part of a Near East and Eastern Mediterranean 'on the edge' of falling under the influence of a pan-Arab agenda which conceived the Mediterranean as part of its territorial project (Figure 4).

Since 1956, Lebanon had a new president of the Republic, Camille Chamoun, known for being openly pro-Western and for advocating liberal capitalism. Chamoun's economic policies exacerbated internal inequalities, especially in the Muslim majority areas of Beirut's periphery (Kassir 2010). Theodor Hanf (2015) saw pre-war Beirut a 'city of fissures [where] the level of integration decreased from the city centre outwards' (2015, 202) towards the suburbs. The newer and peripheral the district, he argued, the less integrated. Hanf described the discrepancies between different groups inhabiting the same city:

> 'The Palestinians disliked Lebanon, where good living had precedence over the liberation of their fatherland. The Shi'i refugees could not reconcile this extravagance with the inability to protect their villages. The Syrians in Beirut had only temporary residence permits anyway. To all of them, this prosperous and cosmopolitan Beirut was a symbol of un-Arab westernisation and decadence' (Hanf 2015, 202).

The nationalization of the Suez Canal in 1956 and the creation of the UAR in 1958 were external events that increased the support for Nasser in Lebanon and made the polarization between Chamoun's pro-Western field

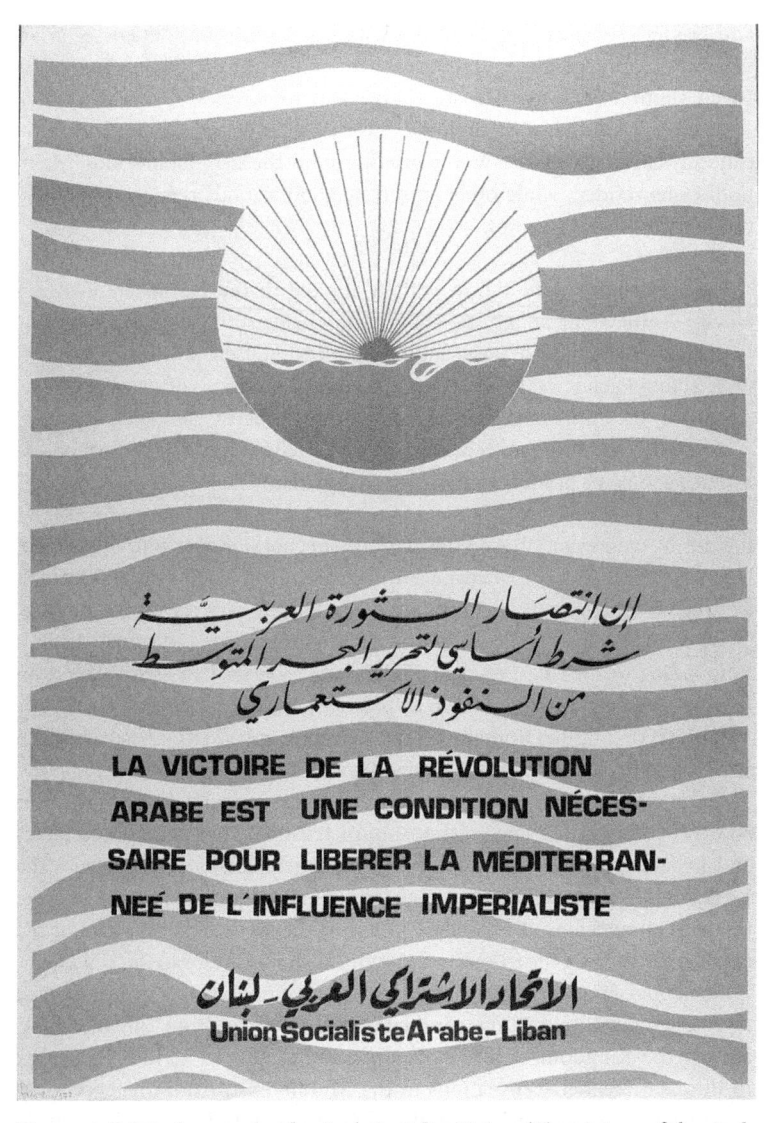

**Figure 4** Political poster by the Arab Socialist Union, 'The victory of the Arab revolution is a necessary condition for freeing the Mediterranean from the imperialist influence'. American University of Beirut, Archives & Special Collections. Political Posters. Poster 177-PCD2081-25.

and Nasser's socialist supporters even starker. Months of politico-sectarian tension culminated in the targeted assassination of opposition journalist Nasib Matni (a Maronite Christian) on 7 May 1958.

Civil war ensued in the following weeks. Mainly perpetrated with low-calibre weapons, violence was focused around the neighbourhoods where politicians resided, while other mixed neighbourhoods remained virtually untouched. As Kassir notes,

> 'The hostilities were almost entirely limited to the perimeter of the city centre, along an arc stretching from Quantari to Gemmayzeh and including Mussaitbeh, Basta and Bashoura but not the downtown itself. [...] the violence was largely restricted to the highest seats of government, notably the Presidential palace on Qantari Hill where Chamoun had dug in, personally firing on rebel forces; to the redoubts of the opposition, such as Salam's headquarters on Musaytbeh [Hill]' (Kassir 455).

It is during these events that President Chamoun, not without strong opposition, invoked the Eisenhower Doctrine and requested help via the US ambassador, alerting him that groups of citizens influenced by Nasser's pan-Arab territorial project as well as by communism were threatening Lebanon's stability. Responding to the call, President Eisenhower mobilized the sixth American fleet in the Eastern Mediterranean and started Operation Blue Bat. On 15 July 1958, 1500 US Marines from the US Sixth Fleet disembarked on Khaldeh beach south of Beirut and brought the situation under control within weeks.

Behind the decision to intervene were specific imaginative geographies about the amicability and relationship of trust between the United States and Lebanon, that complemented the strategic calculus (Little 1996). The United States scripted Lebanon as a place of intervention, where an external threat was menacing it as well as much of the Near East. Explaining the reasons for sending the troops to Lebanon, US ambassador to the UN Henry Cabot Lodge Jr stated: 'The territorial integrity of Lebanon is increasingly threatened by insurrection stimulated and assisted from outside. Plots against the Kingdom of Jordan ... are another sign of serious instability in the relations between nations in the Middle East. And now comes the overthrow ... of the legally established Government of Iraq' (UN Security Council 1958). Lebanon's sovereignty is scripted as something to salvage, to bring back from the brink (of Arab nationalism and of communism), shaping policies of interventions that were not limited to Lebanon, but were echoing across several countries where US and Soviet influence was being contested: 'The events in Iran had a decisive echo in Lebanon, the experience of Guatemala was not lost on Beirut, while Congo was linked by the unfortunate experience of the United Nations in both cases' (Gendzier 1997, 5–6). This moment of US interventionism in what was scripted as a state standing on the fault line between the Soviet and US bloc also relied on specific views of Beirut, its space, and the kind of sovereignty arrangements that became apparent once the troops were on urban ground.

## The urban geopolitics of Operation Blue Bat

Specific urban political economies underpinned what is popularly known as 'Beirut's "Golden age"' of economic growth and political stability in the years preceding Operation Blue Bat. Not only Lebanon was a country of transit for oil pipelines in the Mediterranean and oil revenues from the Gulf (Gendzier 2006), but as Kassir aptly analyses, the city's major transport infrastructures developed as a result of geopolitical changes which intimately linked Beirut to the Gulf oil economies and raised the stakes of US interests in the Mediterranean Levant. The creation of the state of Israel (and the closure of the border with Lebanon after the Arab-Israeli War) and the breakdown of the Lebanese–Syrian customary agreement determined a major increase of Lebanese imports from the United States (in wheat especially). Besides, Beirut's port took over activity from the port of Haifa to serve the Levant as well as the Gulf in the import economy that was thriving from the oil revenues; and Beirut's Khalde airport (opened in April 1954) affirmed itself as a main air transport hub to link the United States and Europe to the Gulf (Kassir 2010, 355–58).

In the US foreign policy representation of Lebanon and the Mediterranean Levant as at risk of penetration from the external threat of communism and Arab nationalism and 'on the brink' of falling into their sphere of influence, a further aspect of the geopolitical scripting of Lebanon links its physical features to its strategic value. In this context, Lebanese sovereignty is scripted as something in particular need of intervention and salvaging because Lebanon is a 'small state'. If the Near East is on the edge, 'tiny' Lebanon is even more so. The Eisenhower administration saw Lebanon as an amicable, albeit tiny spot threatened on all sides: a 'pro-Western island in a sea of Arab nationalism and Soviet subversion' (Little 1996, 25). Intervention was 'in response to an appeal for help from a small and peaceful nation which has long had ties of closest friendship with the United States' (Eisenhower 1958a). The scene at the marine's arrival described by Eisenhower's emissary Robert Murphy and reported on the press of the time expresses the Lebanese amicability towards the US troops who were 'greeted by friendly crowds of late afternoon beachgoers and throngs of peddlers hawking everything from humus to Coca Cola' (Little 1996, 17).

Salvaging the sovereignty of a small amicable state, however, was not simple. The suspicion that Lebanon was as complicated as much as it was amicable haunted the Eisenhower administration before launching Operation Blue Bat (Little 1996). Domestic sovereignty, especially, was a complicated issue that the operation commanders brought directly to the attention of the US Government. Particularly problematic was the constrained role of the national army in Lebanon's multi-religious society, and the fact that any decisive intervention in internal conflict could trigger sectarian divisions. Lebanon's *de facto* sovereignty presented a complexity on the ground which appeared very different from the tiny amicable nation portrayed by official geopolitics, as both territorial control and the monopoly of political violence were not straightforwardly implemented by the state. In Robert Murphy's memoirs, this complexity becomes embodied into a rebellious and hard to access urban fabric through his account of the urban fighting in the inner city quarter of Basta. Murphy tells of his visit to President Chamoun in his residence in during Operation Blue Bat, while the British embassy and the US Marines were also under fire by militants shooting from the dense old city core of Basta. In orientalizing tones, Murphy describes the 'rebellious' Basta as 'a complex of ancient streets and buildings forming the type of district sometimes called the Casbah' (cited in Khalaf 2002b, 118). He continues:

> President Chamoun told me that he had ordered and begged General Fouad Chehab, who was in command of the Lebanese army, to clean out Basta, but without success. My immediate reaction was that Chehab ought to be fired, a competent new commander appointed, and action taken to restore order and authority of the government. I found it was not quite that simple. (Murphy 1964, 399–400, quoted in Khalaf 2001, 118)

In his personal account *Turmoil in Beirut*, former Ambassador Desmond Stewart, who had landed in Beirut on his way back from Iraq, explained the limitations in extending domestic sovereignty to the rebel quarters of the city, due to the confessional nature of the Lebanese social contract:

> *Either the General would have to comb out Moslem and Druze soldiers and send an all-Christian army into Basta, which would split Lebanon for ever, or he would have to send an ordinary mixed detachment into Basta, which would risk a disintegration on the field. (Stewart 1958, 43)*

These domestic sovereignty limitations and Beirut's physical urban layout reinforced each other in creating a script of Lebanon as a complex and at times ungovernable place. In depicting the Marines' arrival, *Life* magazine used orientalist tones to describe Beirut as a lazy, yet potentially treacherous place: 'Beirut was dozing at lunch hour ... when word flickered across the city, "There's a fleet offshore". People were streaming from town to watch the show. Among them were men I had seen at rebel headquarters. One muttered, "Let them come. We are not afraid"'. The article continues describing the contextual ambiguity awaiting the Marines' arrival in the name of the script of salvation '[in the Khalde airfield, to become Beirut International Airport], a marine unslung his 90-pound pack and asked the age-old question, "How the hell do you tell the difference between friends and rebels?"' (*Life* magazine 1958, 15).

This kind of exotic depictions of the city was widespread in popular media accounts of the time. A vignette by British cartoonist Michael Cummings in *The Daily Express* of 9 July 1958 (Figure 5) mocks both the cautious and almost futile character of the intervention, and the laid-back, siesta-prone tiny land of Lebanon. It depicts President Eisenhower aboard a US military ship, its disproportionate mass overshadowing a tiny piece of land, on which a sign reads 'LEBANON. It is forbidden to shoot during the siesta hour'. Looking down on the beach, where a person lounges under an umbrella with written 'U.N.O.' (UN observers), Eisenhower is depicted as

**Figure 5** US Marines intervention in Lebanon. Cartoon by Michael Cummings, *The Daily Express*, 9 July 1958.

signalling to a military and saying, 'Hold everything, Commander – all I can see is a Paraguayan observer firing off a round of whisky and sodas'. Another sign on the beach reads: 'Rebels are requested not to point their guns at anyone' and 'President Chamoun welcomes careful intervention.'

While in 1958 Lebanon's sovereignty was considered as something to salvage through a high-scale response following a request of intervention, quite oppositely in 1975–1976 – as we will see in Chapter 5 – international players considered Lebanon's sovereignty as lost to 'irresponsible' nonstate actors and to 'fatal' mechanisms of chronic violence. In 1975, sovereignty that had been lost due to a deviation from a 'normal' situation of coexistence between confessional communities: a position which, in the eyes of international diplomacy, morally and pragmatically justified nonintervention.

The significance and influence of Operation Blue Bat and its geopolitical script of salvation of a friendly, small and vulnerable, yet confusing nation, did not fade. Quite the opposite, the success of Operation Blue Bat constituted a motor for US intervention in Vietnam. The disastrous aftermath of the Vietnam War, however, and US consequent isolationist foreign policy in subsequent years, in turn contributed to shape the foreign policy of nonintervention during Lebanon's civil war in 1975–1976.

Less than thirty years later, therefore, a very different script shaped the foreign policy of several states besides the United States towards Lebanon. Yet, this should not be seen as a complete rupture, as some aspects of the ways in which the United States scripted Lebanon in 1958, re-emerged in 1975. Alin (1994, 65) remarks how US foreign policy tended to be more preoccupied with state borders in the Middle East than with Lebanon's domestic scene. Differently from 1958, in 1975 the United States spatialized threat in and around Lebanon as unmappable, therefore uncontrollable. Lebanon became a place difficult to map, hence impossible to comprehend and confront. In 1975 the existence of Lebanon itself as a sovereign nation is in doubt. As Paul Salem put it, 'Lebanon was not perceived as being lost to the Soviet Union or to its clients; it was merely being lost to itself' (Salem 1992). Despite relying on and producing specific economic, social and geopolitical rationalities (Fregonese 2009a; Hourani 2010a; Marwan George Rowayheb 2006), the civil war was often portrayed in official foreign policy as total chaos. During its first phases, which were crucial in determining the physical partition of Beirut the United States, oppositely from 1958, scripted Lebanon in ways that legitimized non-intervention in the name of the preservation of Lebanon's sovereignty and territorial integrity.

4

# Towards War

This chapter offers a spatial and urban view on the escalation towards civil strife in the years preceding the start of the civil war in April 1975. It focuses on the spatial and material implications of controversies over security and sovereignty. This view tackles a missing link – already pointed out by El-Khazen (2000, 259) – between the country's positive socio-economic performance in the late 1960s and early 1970s, and the escalation of violence that culminated in war. As we will see, this culmination is discordant from quantitative socio-economic indicators of the time, whereas considerations around unequal development and lack of urban planning (by no means unique to Beirut) do not, on their own, provide a solid argument as to why and how the situation precipitated into open violence.

The first of the geographies I want to explore here is shaped by the tensions over sovereignty that had been brewing between the Lebanese state and the Palestinian armed groups present in Lebanon (and especially the Palestine Liberation Organization, PLO) since at least 1969 (El-Khazen 2000, chapter 12). The second consists of a series of security incidents that shook particularly the cities of Sidon and Beirut. These incidents manifested the deep contest over sovereignty between the state and the Palestinians in and outside the camps, but most importantly, the aftermath of these incidents established on the ground specific *urban architectures of enmity*[1] (Derek Gregory 2004; Lisle 2016; Shapiro 1997). The representation, use, and reorganization of urban territory – of Beirut above all – became the tools for turning 'antagonistic framings' of social and (geo)political difference (Lebanese/Palestinians, Right/Left, Christian/Muslim, and so on) into spatial 'facts', underpinning specific and wanton hostile actions against urban places.

## The missing link: Revisiting socio-economic perspectives on violence in Lebanon

The period between 13 April 1975 and 21 October 1976 is known in Arabic as *al-harb as-sanatayn* (The Two Years' War). Nineteen Lebanese and seven Palestinian armed militias, with varying numbers of fighters and weapons and with sponsorship ramifications in other countries (El-Khazen 2000; see Tables 22.1, 22.4 on pages 300, 302, 303), took to urban warfare for the control of specific portions of Beirut. The Two Years' War was a deadly conflict with enormous human and material costs. In October 1975, only six months into the war, the UK ambassador to Lebanon quoted unofficial estimates that set the damage 'to property, plant and stocks' at £700 million, without including the substantial losses for tourism (Foreign and Commonwealth Office 1975b). There is discrepancy between datasets and official or at least reliable data on the exact quantity of losses and damages are lacking, and often exaggerate or conservative estimates from various sources have had to be adjusted (Labaki and Abou Rjeily 1993, 20–25). However, the loss of human life has been calculated by Labaki and Abou Rjeily (1993) at 19116 and the wounded at 15204.[2] These numbers don't include the kidnapped and the disappeared (170 in 1975, although data for 1976 is not available) and include members of the regular forces (by 1976, 22 per cent of the members of the Lebanese army, Interior Security and General Security had been killed), civilians and militias.

Additionally, about 450,000 residents were forcefully displaced across Lebanon during the Two Years' War, and Beirut gradually saw a good part of its population emigrate abroad, in a country-wide trend that in 1976 brought a negative net migration rate of -297000, meaning that more people were leaving Lebanon than immigrating to it (Labaki and Abou Rjeily 1993, 94). Still, these data are partial: violence was set to continue for twelve more years after a year of military interposition by the Arab Deterrent Force (ADF) in 1977. By the end of the war in 1990, one third of Lebanon's population had been forcefully displaced (Labaki and Abou Rjeily 1993). The material losses of fifteen years of war were huge. The widespread physical destruction included cultural assets such as libraries, education and religious buildings (Labaki and Abou Rjeily 1993), as well as factories, hotels and other industrial facilities, affecting mostly the sectors of industry, tourism, entrepot trade and health (Nasr 1990). While some sector of the Lebanese economy, such as construction, actually continued to thrive (Glasze 2003), the country lost its agricultural and industrial competitiveness on the regional markets, the cost of life increased dramatically, and the general climate of insecurity drove away investment and industrialization (Nasr 1990).

This intensity and spread of violence appear even more unfathomable when we relate them with qualitative and quantitative evidence of Lebanon's glowing socio-economic performance, on the domestic and international scene, at the eve of the war. A telling example of the stark contrast between the country's (and especially its capital city) 'golden age' of economic prosperity and the rapid descent into violence in early 1975 was the substantial construction happening in the areas surrounding the historical city centre. Among these projects was the opening, in 1974, of the *Saint Charles Civic Center*, resulting from a Lebanese and Kuwaiti investment of 20 million Lebanese liras, with a similar projected fixed income for 1977 (Al Hawadess 1975, 85). The project was set to become an urban landmark for dimensions and grandeur, hosting unique amenities (like the Pinnacle rotating restaurant at the top of the Holiday Inn hotel located in the complex) and by mid-1975 (the beginning of the war) it was supposed to 'have on its top a large sign with its name lit in red, each letter [...] 10 meters high' (Al Hawadess 1975, 85). Slightly more than one year from its opening, this enormous investment of capital went from being the absolute landmark of urban growth – not without controversy for the unequal development and infrastructural burdens it created on a city without a master plan – to a notorious landmark in the battle for the determination of Beirut's main dividing line, as we will later see.

While socio-economic factors had a degree of influence on the escalation towards the war, El Khazen (2000, 261–62) argues, however, that there is no direct causation between these and the breakout of civil war. In macro-economic terms, Lebanon was steadily growing. The national GDP increased in all sectors across the 1960s and until 1974 and – bar a decrease in agriculture's share of the GDP[3] – the total GDP triplicated from 1961 to 1973 (Labaki and Abou Rjeily 1993, 177). Finally, with the increase of the minimum wage in 1974, equality of income distribution had improved (Labaki and Abu Rjeily p. 178).

It is worth looking beyond quantitative indicators, and at the work of sociologists and anthropologists especially, to appreciate the complex factors that contributed to shape deep and long-term discrepancies alongside the country's apparent economic stability between Independence and 1975.

As eminently analysed by anthropologist Souad Joseph, the sectarian affiliation of individuals and institutions heavily mediated state resources allocation and political party services (Joseph 1975). Rather than a mere 'descent into barbarism' as portrayed in the press and in diplomatic discourses (as we will see in Chapter 5), if anything, the war acted as normalizer and amplifier of the pre-war dynamics analysed by Joseph. Indeed, dynamics incepted before and crystallized during the war, have become established nowadays. Sectarian affiliations and institutions are integrated in daily practices of service provision, security, resource allocation and infrastructure works (Bou Akar 2018; Fawaz, Harb, and Gharbieh 2012; Joseph 1983, 1975; Monroe 2016; Nucho 2016). During the war, resource and service provision, including electricity to garbage collection (Harik 1994), interlocked with sectarian affiliation up to the point of becoming spatially manifest in Beirut through the purge of sectarian diversity from neighbourhoods and the partition of the city. Moreover, in her 1983 work on Camp Trad (Joseph 1983), Joseph has uniquely brought together reflections on the politics of sectarianism, the class and spatial implications of inter-sectarian residents' projects (often in less affluent neighbourhoods), with the geographies of wanton violence in the civil war. She argued that planned attacks against mixed and Muslim areas in the east of Beirut – often perpetrated by militias affiliated with the Christian ruling class – aimed to undermine exactly those intersectarian relationships that menaced the sectarian establishment. Here, exactly that possibility of heterogeneity that Martin Coward identifies as the target of urbicide (Coward 2009) is outlined by Joseph as the target of a struggle for power and resources that was at once sectarian and about class inequality.

In other words, even what is referred to as the country's 'Golden Age' (Kassir 2010), when particularly 'Beirut came to radiate prosperity to such an extent that hardly any part of the country was left entirely untouched by its influence' (Salibi 2003, 191) contained deep societal discrepancies that translated spatially. The newer and more peripheral the district, Hanf (2015) argued, the less integrated: 'Whereas the old Beirut had long experience of coexistence, most of the inhabitants of the newer districts had moved from homogeneous villages to an equally homogeneous urban environment' (Hanf 2015, 202).

Fouad Shehab's Presidential term between 1958 and 1964 was marked by attempts to instate a sense of national unity through state-driven regional development and planning. This included national education and welfare policies, as well as the inception of a range of centralized institutions including a Ministry of Planning and an Executive Board of Public Works. Shehab's term was followed by the very different Charles Helou's presidency (1964–1970), characterized by a more liberal economic policy oriented towards increasing the financial sector and the service economy. By the late 1960s, then, Beirut had become the core of economic, administrative, and cultural life, leaving a noticeable developmental gap between itself and the rest of the country, especially the agricultural south. This phenomenon – also known '*Macrocéphalie Beyrouthine*' (Bourgey 1985, 11) – indicated the demographic and functional concentration in Beirut compared with the rest of the Lebanese territory. The physical appearance of Beirut was also changing fast, as real estate speculation took up pace and Beirut's red-tiled roof skyline was gradually turning into a cluster of modern towers, while the so-called 'misery belts' continued to sprawl in the peripheries, a constellation of lower income areas and Palestinian refugee camps (Khalaf 1993; Boudjikanian-Keuroghlian 1994).

The spatial inequalities in and around Lebanon's capital were all too clear at the time, as denounced in the magazine Al Hawadess (Shukrallah Haidar 1975):

> Lebanon is suffering from several issues because the growth and development have occurred outside any official and responsible planning. Some of the symptoms of this cancerous growth, and some of the issues that Beirut is facing, are the residential slums that surround the city and suffocate it, the traffic congestions, the narrow streets, the lack of pavements that can help reduce pedestrian traffic on the roads, and pollution and the absence of any areas that allow for the city and its inhabitants to breathe! The city's infrastructure is basically unable to keep up with the city's rapid growth [...] The government [...] tried to resolve some of these issues but their approach lacked the main ingredient that would have provided a natural and comprehensive solution: a large scale urban and civil plan for the city.[4]

Still, the planning perspective alone does not explain how these spatial fault lines became violent divides and what the role of wider geopolitical issues was in the process. The socio-economic indicators are also partial in explaining violence, first because these tend to look quantitatively at Lebanon's performance on the international scale and second, as they tend to equate positive economic performance with social improvements. Besides, Lebanon's service-driven economy and the laissez-faire attitude of the state to economic affairs were not new phenomena at the eve of the war (Farid El-Khazen 2000) and, although there were issues with unequal income distribution and uneven development, including insufficient government investment in rural areas and insufficient measures to cope with rural exodus towards Beirut's suburbs (el Khazen, 262), 'the ills associated with Lebanon's socio-economic development since independence in the 1940s did not suddenly emerge in the mid-1970s' (Farid El-Khazen 2000, 260).

## Hybridizing sovereignty: The relationship between state and irregular armed groups

A spatial account of the evolving relationship between the Lebanese state and the political and armed role of the Palestinian groups and especially the

Palestine Liberation Organization (PLO) is crucial to establish the missing links and redress the apparent discrepancy between Lebanon's positive socio-economic outlook, the presence of spatial and economic inequalities which were, however, unexceptional, and the country's rapid descent into internecine violence.

In 1948, 100,000 Palestinians crossed into Lebanon to escape the first Arab–Israel war. According to the United Nations Relief and Works Agency (UNRWA), today the descendants of the 1948 exodus amount to 455,000 Palestinian refugees registered in Lebanon today. While Palestinians of the urban upper and middle classes and most Christian Palestinians resettled in Lebanese cities, most refugees were settled in sixteen refugee camps across the country, mostly clustered around the edges of Beirut, Tripoli in the north, and Sidon and Tyre in the south of the country. Of those sixteen camps, four were to be destroyed during the Two Years' War.

The first significant episode in the changing spatial relation between camps and state occurred during Fouad Chehab's presidency and period of national social reforms. Already the first commander of the national army after the end of the French mandate (Dīb 2006) and in charge of managing the 1958 crisis (Chapter 3), Chehab was elected President of the Republic later in 1958 and, until the end of his presidential term in 1964, he dealt 'with the country's political problems [through] a more balanced regional development, the building of an efficient state bureaucracy, and the launching of long-term planning' (El-Khazen 2000, 177). Shehab's public policy aimed at creating a higher sense of national belonging by homogenizing the development and the distribution of wealth not only in Beirut, but throughout all the Lebanese territory, including the neglected rural areas where the rural exodus was already under way. Shehab was keen on projecting an image of Lebanon on the world geopolitical map – then dominated by the Cold War dual blocs reasoning – as a neutral nation, neither openly pro-Western nor openly pan-Arabist (El-Khazen 2000). In order to establish Lebanon's neutral geopolitical role, stability had to reign at home. And with Palestinian political and armed activity increasing in the country since at least the 1958 crisis (Barak 2009), the Palestinian refugees and their camps were the sites where the price for domestic stability had to be paid.

As the second decade of Palestinian presence in Lebanon began, Chehab's focus on neutrality and stability translated into increased political control by state authorities in the camps. The Deuxieme Bureau, the Lebanese army intelligence section, was present in every camp, working alongside the Lebanese police, the Directorate of Refugee Affairs, and UNRWA-designated camp leaders. The Bureau sought to suppress all forms of Palestinian political activity and organization, vetting UNRWA employees and appointments, recruiting spies and informants in the camps, and intimidating, arresting, beating, and even torturing activists in some cases (Sayigh 1994, 1977). State repression extended further, materialized in the built environment of the camp and its shelters: 'roofs were not allowed, cement was prohibited material, cartographic boundaries were rigorously enforced' (Abourahme 2015, 207). These state activities – 'nearly a decade of intimidation and extortion' (Sayigh 1994, 68) – produced deep resentment among the Palestinian population.[5]

If on the one hand, Fouad Chehab had been strengthening national unity and sovereignty by encompassing the whole territory of Lebanon under a set of social reforms, and neutrality abroad by ensuring domestic stability, on the other hand this unity and stability had to rest on the assurance that the Palestinian population would remain other from the Lebanese, and that their spaces, the camps, would remain separate from the Lebanese nation and body politic. In other words, the unity and neutrality of Lebanon were structured around an architecture of enmity which aimed to curb Palestinian agendas and differentiate the Palestinian population – spatially, socially, and politically – from the Lebanese one. Strengthening the national unity and Lebanon's geopolitical agenda abroad, equated to strengthening the power grip on the Palestinians at home.

The second episode was the stipulation, in November 1969, of the secret Cairo Agreement between the chief of the Lebanese army Emile Bustani and PLO chief Yasser Arafat, secretly signed in Cairo in the presence of Egyptian president Jamal Abd el Nasser. The agreement was ratified by the Lebanese Parliament later in the same year, and effectively sanctioned the transfer of security and authority in the Palestinian refugee camps from the Lebanese Armed Forces and the *Deuxième Bureau* to the Palestinian Armed Struggle Command and it also granted freedom of movement to the Palestinian resistance in the Arqub region of south Lebanon, where a network of military bases and supply trails had been built since at least 1967 and where the local population's grievances against the colonial mandate border which since 1920 had divided lands and towns within Galilee, were channelled into solidarity with the Palestinian resistance (Kassir 2010, 474).

The Cairo agreement was a response to the evolving power and spread of support (both paramilitary and cultural) for the Palestinian resistance outside of the refugee camps and into the Lebanese urban space. Over the course of several major security incidents and ensuing confrontations between the army and the resistance, in 1968 and 1969, it became clear that the Lebanese regular forces could not contain the growing presence of the Palestinian resistance, which by now included Lebanese armed supporters, without risking a civil war. There was also increasing public support for the resistance by Lebanese officials, especially by the Prime Minister, an element which contributed to an atmosphere of growing rivalry and deadlock between the two highest state powers (the PM and the President) which led to then PM Rashid Karame to resign in refusal to launch a military operation against the resistance on order of President Charles Helou in April 1969. The Cairo Agreement became the response to the political deadlock, as an agreement with the resistances became the precondition for the PM to serve again (Kassir 2010). One further major parallel dynamic was the expulsion of the PLO from Jordan following the actions by the 'Black September' group in Amman in 1972 and the relocation of numerous Palestinian guerrillas from Jordan to Beirut and the consequent installation of the PLO headquarters in Beirut. These tectonic shifts in Palestinian political geography at the end of the 1960s and early 1970s provoked remarkable internal tension in Lebanon: the Palestinian guerrillas often clashed with the Lebanese security forces, and the political panorama was more and more divided regarding the Cairo Agreement and the position of the army vis-a-vis the Palestinian armed groups, as well as on the wider nature of the Palestinian armed presence within Lebanese territory.

The era of the Palestinian Revolution was an intense and fraught one, where growing Palestinian militant activity 'exacerbated the contradictions of the Lebanese system' (Brynen 1989, 50), made Lebanon increasingly unstable and blurred the boundaries of sovereignty and security between the camps and the outside. The PLO was able to build a huge power base in Lebanon, controlling the camps officially and well beyond the boundaries of the camps unofficially. The southeastern Arkoub district, where guerrilla activities were focused, became known as 'Fatah Land' (Khalaf 2002a, 217), and the PLO functioned effectively as a 'statewithin- a-state' (Sayigh 1997, 21). This 'ministate' employed as much as 65 per cent of the Palestinian workforce and offered free medical services, education subsidies, and camp infrastructure (Sayigh 1994, 213). Palestinian movements cultivated alliances with Lebanese parties within and outside government, which gave the latter access to both financial patronage and arms supplies from the PLO and their sponsors (Hanf 2015). Armed raids by Palestinian groups against Israel resulted in repeated Israeli military interventions in Lebanon, and Lebanese civilians in the south – many of whom had supported and joined the Palestinian movements initially – were increasingly caught in the crossfire. Attempts by the Lebanese army to rein in Palestinian groups resulted in further clashes in 1973. In one such episode in April 1973, Lebanese National Movement militias fought alongside Palestinians against the army, which was bombarding the camps around Beirut (Turki 1988). Even the Lebanese army became divided, with commander Ahmad Khatib threatening to shoot soldiers if they fired at Palestinians; Khatib and his troops later did split from the army command in the first year of the civil war.[6]

These events pointed not only to the polarizing nature of the Palestinian question in Lebanon, but also to the further *hybridization* of sovereignty. While the Lebanese state's ability to confront Palestinian armed groups decreased, a growing number of political groups, occasionally with the collaboration of members of the army itself, recruited and armed their own irregular militia forces. And this brings us to the third episode in the hybridization of sovereignty in Lebanon ahead of the war, which was the increasing armament of Lebanese political parties and the paramilitary spaces and practices developing in and around Beirut. The Cairo Agreement had essentially granted liberty to the Palestinian guerrillas to conduct anti-Israeli resistance operations from the Palestinian camps in Lebanon, in which the PLO and its militias now owned a high degree of territorial and logistic autonomy. Although this liberty was formally limited to the perimeter of the camps, the guerrillas had started to assert their presence inside cities, occasionally even by instituting roadblocks and performing displays of weapons. The Lebanese army came under the spotlight by the Lebanese political right as unable to control the Palestinian resistance. This happened in a situation where virtually all political parties 'began to acquire heavy weapons and to engage in organized military training' (Farid El-Khazen 2000, 225) facilitated by Al *Tanzim* (the organization). Defined by Hanf as 'not so much a fighting force as an organization for training guerrillas' (Hanf 2015, 191), this was a smaller and secret 'elite' unit, funded soon after the Cairo Agreement, and headed by Fouad Chemali (El-Khazen 2000, 225), providing basic military training to young armed supporters 'with the help of sympathisers within the army' (Hanf 2015, 191), and 'including the highest levels of both the general staff and the intelligence services' (Kassir 2010, 507).[7]

One of the former fighters I interviewed, Edouard trained in the Lebanese Kata'ib (*Phalanges*) militia and recalled how armed party supporters used the mountainous areas of Keserwan north of Beirut as training grounds to prepare to counteract the Palestinian resistance in the cities:

> [Al-Tanzim] tried to gather young Christians, mostly, who were against the Palestinian occupation of the country. Starting from 1973, the Christian parties, essentially the Kata'ib, began to train their men [sic] for combat [...]. There were secret trainings, in certain regions of the country, where people – of which I am one – spent weekends training instead of going dancing. They spent the weekends training, preparing, because there will be one day in which we will be brought to fight the Palestinians. (Edouard)

Nizar, a member of the militia coalition known as *Harakah al Wataniyya* (National Movement), also testifies the practice of many parties training to use weapons:

> When I was studying [...] there were people [with whom] I was not directly connected, but who were [...] from my neighbourhood, who had already had confrontations with each other. They were saying that they were already carrying weapons and training to fight [...]. [Among these] there were people both from Kata'ib and from Ahrar. (Nizar, 1 December 2005)

The race to arms was not only among militants of the Lebanese right. Albeit in a less organized manner and with initially less resources, the supporters of a number of parties close to the PLO were also provided with training and weapons by the latter (Kassir 2010).

Between 1958 and 1974, the line separating de jure state sovereignty (the government, the army) from non-state and paramilitary (PLO, militias, Tanzim) de facto sovereign practices blurred and the power relation between the Lebanese state and the armed Palestinian presence inside Lebanon was irreparably undermined.

These dynamics contributed to shape a number of urban architectures of enmity. First, since 1958 Fouad Chehab's redesign of Lebanon's geopolitical role as a neutral nation suppressed Palestinian political agency and drew a political geography that distinguishing starkly between the space of the camps and that of the city and established those distinctions as an indication of sovereign stability. Second, in 1969, the Cairo Agreement unravelled the Chehabist design, blurred the political geographies of distinction between camp and city, culminating in the 1972 PLO HQ establishment in Beirut. With the relationship between the army and the Palestinian resistance compromised, mistrust at political and popular levels rose. Third, the increasing armament of militias both on the right and the left of the political spectrum, and the partaking of elements of the army in the training of militia members, established the physical (training grounds) and material (weapons) premises for war. These architectures took shape in ways that no socio-economic indicator could isolate and identify, and no unequal development thesis could, alone, explain. Underpinning these architectures, was a deepening fundamental rift within the idea of Lebanon as a nation and its role in the world.

## Urban architectures of enmity

The unravelling of the geographical and sovereignty distinction between camp and city, and the establishment of the physical and material premises for violence rest on a wider sovereignty rift, where 'the Palestinian logic of the struggle against Israel could not be reconciled with Lebanese considerations of state' (Hanf 2015, 171). This rift cuts straight to the core of the sovereignty question of who controls territory and political violence; and of Lebanon's role and borders vis-à-vis the wider geopolitical map of the Arab world. In other words, 'whereas the growth of militias among the Christians had proceeded from a perceived need to protect national sovereignty, among the Muslims it was justified in terms of the "defense of the Palestinian Revolution an o the Arabness of Lebanon"' (Kassir 2010, 509).

It is within these deep, long-gestating architectures of enmity and their geopolitical connections that we ought to contextualize two specific security incidents in Sidon and Beirut, respectively in February and April 1975. These very serious events revealed the urban political geographies that the Cairo agreement had produced: the blurring of the distinction between camp and city, and – most importantly – the polarization along political and religious lines on the urban ground.

## The Sidon fishing protests

The incidents that occurred in Sidon on 27 February 1975 are considered as a prologue (Kassir 2010) to the civil war. The fishermen in the southern port of Sidon were protesting against the concession for deep-sea fishing off the coast to a Sidon-based company called *Protein*, which they accused of threatening their livelihoods, and of which President Camille Chamoun acted as chairman (Hanf 2015). On that day, Marouf Saad, the Sidon representative in the Lebanese parliament and founder of the Popular Nasserist Organization, was walking at the head of the march. As the protest turned violent, Saad was shot and died in hospital in Beirut a week later. There have been several attempts to outline a coherent hypothesis about the assassination, and diverse interpretations of the circumstances demonstration. In this study, it is important to underline what these socio-politically charged events meant for the control of the physical space of the city of Sidon, as what the rapid takeover of control of key urban nodes meant for Lebanese sovereignty and as a harbinger to read the polarization that the country was by now engulfed in.

Even before Saad was shot, the PLO power network was already embedded within the urban infrastructure of Sidon and indeed competed against Saad's influence. Saad was considered by his fellow citizens and the Lebanese political scene as a local politician working closely with the people; in the 1950s he worked to bring Sidon – a city of the south outside the ray of Beirut's urbanization – within the orbit of President Chehab's program of social reform and territorial development. Saad became za'im (mayor) of

Sidon in the 1960s, thanks mainly to the support of a urban network of Sunni political consensus in the city's old neighbourhoods (El Khazen 2000). Nevertheless, these quarters soon became the terrain of a political power struggle between Saad and the PLO which, especially after the Cairo Agreement, had gathered support and legitimacy beyond the camps, often as a backfiring of the Chehabist use of 'Sunni strongmen in the cities to supervise the Palestinians' (Barak 2009, 83). Saad's Sunni power basis and its spatial reflection, then, changed noticeably with the PLO's growing presence in the city, which lies near the two biggest refugee camps in the country: Ain el Helwe and Mieh Mieh. As the Syrian-backed and pro-Palestinian militia Saiqa and the Fatah party opened offices outside the camps and into the city, the competition and polarization between Saad's sphere of influence (the inner city core) and the PLO sphere (the camps) increased until the Palestinian resistance clashed with Saad's Tanzīm al Quwwah al Sha'biiya fi Sayda (Organization of the Popular Force in Sidon) in 1970, when the militias held Saad prisoner inside the camps. What was felt as a humiliating treatment for Saad by the Palestinian resistances, began to be felt publicly as an affront to the city and its sovereignty, and the incident even acquired international tones when Egypt's Nasser dispatched mediators to solve the issue (El Khazen 2000, 277). Eventually, Saad lost his municipal role in 1973.

Immediately after the shooting, PLO armed men entered Sidon's streets and quickly sealed the city off from the rest of Lebanon by blocking the coastal road and instituting check points. Clashes between the Palestinians and the Lebanese security forces followed until, four days later, the city was eventually reopened. By sealing the city off, the PLO stated its territorial presence and power. Sidon's material fabric – its neighbourhoods, its buildings and its thoroughfares – became the strategic terrain where a non-state armed actor exerted their power beyond the zones that the Cairo Agreement had designated for their political activity. The Sidon events had established on the ground the presence and control of the Palestinian resistance and their supporters among the Lebanese (and Muslim). Within the context of widespread radicalization, armament and training of party supporters across the political spectrum, and of increasing paralysis of the army (Barak 2009) for fear of communitarian split, the Sidon events were also an instance in which the deep polarization along the architectures of enmity outlined above became manifest in the urban space: 'Palestinians [...] and the entire Sunni establishment were ranged together for the first time, albeit for different reasons, against the "Christian" army and the parties of the Christian communities. The "Protein Affair" had forged for the Palestinians the broadest alliance they could possibly conceive in Lebanon' (Hanf 2015, 174).

## The bus shooting and the start of the war

The bus shooting that happened around midday on Sunday 13 April 1975 in the Beirut's south-eastern suburb of Ayn al Rummana, is considered the igniting incident of the civil war (Hanf 2015; Kassir 2010) and, similarly to the Sidon incident, the exact sequence of events is still debated. A bus transporting Palestinians returning from a commemoration in the Sabra refugee camp in the south of the city to the camp of Tall al Zaatar in the north east was caught in a gunfire ambush and twenty-seven of its passengers lost their lives. The attack is thought to have been a response to another shooting that had taken place in the same area a few hours earlier. Here, a car – tracked down as belonging to the Democratic Front for the Liberation of Palestine (Hanf 2015, 204) – did not stop at a roadblock and opened fire against a church congregation, where Pierre Gemayel, the leader of the Kataib party was attending a service. In a matter of hours after the bus shooting, fighting spread through the suburbs and in the Palestinian camps of Dekhwaneh and Tall el Zatar, leaving forty dead (Chami 2005). Shooting then continued, particularly intensely, between the two neighbourhoods of al Shiyyah and Ayn al Rummana near the location of the bus shooting.

As narrating Lebanon's violent past remains a highly contested process (Haugbølle 2010; Larkin 2012; Mermier and Varin 2010), there is no commonly agreed version of the events of 13 April 1975. Some elements still remain unclear, such as the unusual route taken by the driver on the day (El-Khazen 2000) as 'for years it had been the custom of Palestinian transports between camps to the east and west of Beirut to make a detour around Ain al Rummaneh' (Hanf 2015, 204 Note 41). The difference of interpretation of the events still resonates in the testimonies of former fighters Edouard and Nizar, who witnessed the immediate aftermath of the shooting. Each frames the event according to different memory cultures (Haugbølle 2010).

> Until nowadays, nobody understood why this bus passed through the streets of Ayn al Rummana and more specifically next to the church where there had been the assassination of Abu Asim[8]: at that moment, in that part of Ayn al Rummana, the Kataib members were furious and they all had weapons, and suddenly a bus appears, transporting armed Palestinians. They gave several explanations, but I know that if I put myself in the shoes of those who were there [...] I would not hesitate to shoot. [...] Now people give an enormous number of interpretations [...] after 30 years, if you look back you can actually say that it was more likely to be set up than accidental. (Edouard)

> We [...] got on the bus that was going down to the centre, and we entered [...] between al Shiyyah and Ayn al Rummana [...] the only separation was a road. We never thought that it [the shooting] could happen. In the end we are children of the same country. So we cut across, and we found that they had shot at a bus. Then, we got off and the driver then said 'Everybody go home'. At first, we were not aware of what had happened, we only knew that they shot a bus which was carrying Palestinians and that the responsible [actor] for the area was the Kata'ib party. (Nizar)

While some scholars argue that the 'site of the incident was of little importance. The shooting could have taken place in another part of the city [...] for in the mid-1970s there was no dearth of sites or pretexts for violence' (El-Khazen 2000, 286), others have exceptionalized the event, associating its location within the sectarian geographies of the city, notably, the location of 'Christian' Ayn al Rummana in a 'Muslim' surrounding of low-income neighbourhoods and Palestinian camps (Sarkis 1993). Both these interpretations are limiting, however. On the one hand, they reduce the geography of Ayn al Rummana and the incident of the bus to a passive and even interchangeable background for violence to unfold: this view erases the urban architectures of enmity that had been developing and layering onto the city in the previous years. On the other hand, they exceptionalize the event in ways that essentialize sectarian identity and deny the contextual and spatial specifications of the event. Instead, the next chapters place the physical space of the city front stage, first for considering the architectures of enmity that shaped the geographies of violence, and second, to understand how urban space shaped 'unofficial' geopolitical imaginations about the idea and role of the Lebanese nation. These narratives were crucial in the establishment of a new political map of Beirut and the violence that this map underpinned.

## The Two Years' War

In the week following the shooting, fighting reached the coastal towns of Tripoli, Sidon and Tyre, until the secretary-general of the Arab League Mahmood Riad proclaimed a ceasefire that lasted four days on 16 May. Two other *jawlāt* ('rounds' of fighting) followed, between 20 and 26 May and between 30 May and 30 June. Battles took place also in the western parts of Beirut, in the coastal town of Damour and in the north of the country. The fighting in the first weeks of the war was mainly confined to the suburban areas of the capital, and daily routine in the city centre went on almost uninterrupted. Every round of fighting, however, would carry the harbingers of worse things to come, shaping a spiralling pattern of *round-ceasefire-longer round* to the early phases of the war, adding to the atmosphere of tension in the city:

> In the first round the fights were a few [sporadic], while in the second round they became more, the first round they would last 2 or 3 days and then a truce would come, for a week or 10 days, but the atmosphere was charged. The following rounds were always was wider than the ones before. For example, the second round would last 2 weeks and after that a truce would come, and after one week or 10 days another round was longer. (Rashid)

Nevertheless, in late September 1975, the centre of Beirut for the first time became the theatre of militia fighting and of territorial partition as the Christian and so-called 'isolationist'[9] militias, guided by the Lebanese Phalanges, established their presence in the part of the centre overlooking the harbour, which they renamed 'Fourth Sector' (*al qita' al Ruba'*). During the following month, the Palestinian and the Lebanese militias reunited in the National Movement (Harakah al Wataniyya) and began repositioning themselves in Kantari, a neighbourhood on the hills west of the city centre where the high-rise towers and hotels were located. On 27 October 1975, what is known as the 'battle of the hotels' began between the Phalanges and the Al Murabitun militia, with the former positioned in the Holiday Inn Hotel in an attempt to break into the area controlled by the National Movement, and the latter taking over the Murr Tower (*Burj al-Murr*) in the vicinity of the hotel. Amidst diplomatic mediation and the continuation of Israeli air raids against the Palestinian resistance in the southern areas of the country, faith-based and territorial violence in Beirut culminated on 6 December 1975 with the events remembered as 'Black Saturday'. During that day, the killing of four members of the Phalanges triggered a reprisal including the random killing of Muslim civilians in the streets of the city centre and in a series of intense rounds of combat that lasted one week before a ceasefire was reached and the Murr Tower and the Holiday Inn were evacuated in mid-December.

Some of the deadliest clashes between the Lebanese Front and the National Movement took place in the beginning of 1976, as the Palestinian refugee camps of Tell al-Zatar and Jisr al-Basha in the east of Beirut were put under siege by the Katae'b militia. The National Movement's reprisal consisted of attacking the neighbourhood of Horsh Tabet at the southern limits of Beirut municipal boundaries. The attack-reprisal pattern was repeated in the following days when, in reply to the Phalanges' massacre of the residents in the Karantina quarter next to Beirut's harbour, the National Movement forces targeted the Christian population of the town of Damour, south of Beirut. A ceasefire encouraged by Syria was implemented and the deployment of the Lebanese army in the guerrilla hotspots continued, but what has been called 'the war of the barracks' began as Lieutenant Ahmed Khatib formed his 'Army of Arab Lebanon' in the Bekaa valley in the east of the country. Shortly after another army general, Aziz Ahdab, executed a military coup urging President of the Republic Sleiman Frangieh to resign, and his action was soon joined by Lieutenant Ahmed Khatib. Between 21 and 23 March 1976, what has been called the 'Battle of Beirut' or the 'Battle of the Hotels' (*Ma'raka al-Fanadiq*) culminated in the Kantari neighbourhood around the hotels, where after six months of confrontation, National Movement militias took over the Holiday Inn from the Phalanges fighters that were barricaded inside. The aim of the battle was to gain as much territory in the centre of Beirut as possible, but the modalities of the violence were much more complex than a quest for territory: neither a passive object serving tactical aims, nor a symbol of identity existing a priori, the built environment was part of the active, daily and contested production of the geopolitical imaginations of the militias.

Until May the fighting continued across the whole country and in Beirut, amidst the attempts of installing a truce and despite the signature on 14 February 1976 of the Damascus Agreement, a document that was aimed at ending hostilities and at reinstating the integrity of Lebanon in avoidance of the internationalization of the conflict. In the summer of 1976 isolation and scarcity afflicted Beirut as the airport closed down, the power supply was cut due to the damage caused by the fighting and water and medicines were lacking, while more and more foreigners evacuated and many Lebanese sought escape by sea. Attacks and reprisals went on involving both the Palestinian camps and the cities, and in July 1976 the Palestinian camps of Tell el-Zatar fell because of the Phalanges' siege that had been occurring for almost two months.

It was only between October and November 1976 that, in the course of two meetings in Riyadh and Cairo, PLO chief Yasser Arafat suggested a solution to the conflict that was then deliberated by the representatives of six member states of the Arab League – Syria, Egypt, Lebanon, Saudi Arabia and Kuwait – as well as by the PLO. The two key points of the solution to the conflict were the implementation of the Cairo Agreement (see Chapter 3), to be supervised by a committee of four Arab states, and the economic and military engagement of the Arab League in Lebanon. The latter was translated into the institution of an Arab Deterrent Force (ADF), a military contingent under the guide of the Lebanese president, Elias Sarkis, and composed mainly of Sudanese and Syrian soldiers, to be deployed across the whole Lebanese territory. This formula was put into practice through the ceasefire that started on 21 October 1976 and through another meeting in Cairo. Between 10 and 15 November, the ADF entered Lebanon and Beirut from the mountains in the north of the capital and subsequently deployed through the city. The deployment consisted of thirty thousand soldiers – mostly Syrian but also Libyan, Saudi and Sudanese. The Two Years' War ended. Towards the end of November the airport reopened and in December a new government was formed by Prime Minister Salim el Hoss on request of the President of the Republic Elias Sarkis.

# 5

# Lebanon Lost: The Urban Impact of Non-Intervention

*Lebanon's present importance is not great, but a change to an anti-Western regime would harm our interests. We should help to maintain the status quo. (Valedictory despatch on Lebanon of Mr P H G Wright, 28 April 1975[1])*

In the first months of the Two Years' war, as seen in the previous chapter, Beirut suffered heavy life and material losses. *Time* magazine (1975) described the dire economic consequences of the exodus from Beirut, mainly the substantial flow of capital and investment towards other cities such as nearby Athens:

> For decades Beirut had been a magnet for banks and big corporations, a city that took pride and profit in the swarms of glittery limousines that kept its avenues constantly clogged, the fashionable boutiques, the expensive nightclubs and casinos, the whole ambience of opulence and sin. But no one wants to invest or play in a battlefield. For weeks now a full-scale exodus has been under way, and many U.S., British, French, German and Japanese firms are relocating in Athens, an hour from Beirut by air. (Time Magazine 1975)

This is without counting the thousands who, beyond the world of business and often from low-income classes, did not or could not migrate, or did so only temporarily. UK Minister of State for Foreign and Commonwealth Affairs David Ennals, requesting humanitarian aid to the Minister of Overseas development Reg Prentice on behalf of the UK ambassador to Lebanon, explained that

> While it is true that the per capita income in Lebanon is or at least was relatively high, the distribution of income is very uneven and many of those who are suffering most are the poorer Moslems [sic] who have often lost their houses and had their possessions looted or destroyed. Lower middle class and working class residential areas have been badly hit by the fighting

and it is these people, who can least afford it, who have suffered most of the damage and casualties. (Foreign and Commonwealth Office 1975b)

The war-induced exodus from Beirut was the latest addition to a wider history of diaspora from Lebanon (Albert Hourani 1992). This loss of population and capital is paralleled by the loss of the city's cosmopolitan urban landscape. A French reporter is quoted at the beginning of Edward Said's *Orientalism* describing the city as the last vestige of the 'Orient of Chateaubriand and Nerval' being 'gutted' by destruction and division (Said 1995, New:1), the last embodiment of a European Orientalist dream of 'a place of romance, exotic beings, haunting memories and landscapes, remarkable experiences' (Said 1995, New:1) turned into a dangerous urban battleground. This depiction was echoed by representations of the city in popular fiction, such as the war reporter protagonist of Volker Schlöndorff's 1981 movie *Die Fälschung* (in English: *Circle of Deceit*) who, against an apocalyptic urban background of burning buildings, bullets flying overhead, and groups of combatants running back and forth, states: 'the once Switzerland of the Orient is now a no-man's land' (Schlöndorff 1981).

The widespread narrative of loss, as this chapter illustrates, extended to international diplomacy and the discourses it produced about Lebanon and its sovereignty. The country was represented geopolitically, by the foreign policy establishments of several states[2] who had a stake in the conflict, as *lost* in a war deemed tragic, but also incomprehensible and therefore difficult to intervene in. This geopolitical narrative of loss, however, was not limited to diplomatic statements and foreign policy speeches. Instead, it shaped a very concrete policy of non-intervention during 1975 and 1976. While unsurprisingly justified by the laws of non-interference in the sovereignty of a state, this reluctance to intervene on the ground in these early phases of the civil war – mostly consisting of close-quarter urban fighting, purging of neighbourhoods and takeover of urban services by the militias – however coincided with and, I argue, contributed to the irreversible partition of Beirut.

The international attitude towards Lebanon in the early 1970s was substantially different from that of 1958, when (as seen in Chapter 3) the US Sixth Fleet swiftly disembarked the marines on Beirut's Khalde shores in the name of the Eisenhower Doctrine to 'salvage' a nation 'on the brink' of socialism and pan-Arabism. In contrast, no similar territorial intervention took place in 1975–1976 to tackle a problem that was instead represented by Western international diplomacy as ultimately domestic and limited to the territory of Lebanon. In 1975 and 1976, the practice of foreign policy focused on managing the risk of a regional war and rested its case on the fact that the conflict was too complex to intervene. While in 1958 the US troops entered the heart of Lebanon's capital, in 1975 it was 'not a simple question of pushing troops into Lebanon' (White House 1976c), reminded Henry Kissinger – then US Secretary of State– to the members of the National Security Council in 1976.

What was different in 1975 compared with 1958? Firstly, as we will see, US non-intervention was paralleled by a higher degree of confusion over the complexity of the civil war (Salem 1992). Secondly, attitudes to interventionism had shifted due to the US internal situation at a time of anti-war activism against the great losses in the Vietnam 'quagmire'. Regionally, at least three factors influenced US non-interventionism: the 1973 oil crisis and consequently Saudi Arabia becoming a regional power; the consequent reshuffle of US/USSR regional alliances, in particular Sadat's Egypt opening to the United States and eventually contributing to the 1979 Egypt–Israel peace treaty; and the phase of Détente in the Cold War that had started in 1969. Within this situation, the US priority was now maintaining a balance between Egypt, Syria, Israel and Jordan and, under Kissinger, boosting the Israel–Palestine peace process. This situation shifted US foreign policy away from intervention at least until 1983, when Hezbollah's suicide attack against the US Marines' base of the Multi-National Peacekeeping Force led the then US President Ronald Reagan declare that:

> there was a time when our national security was based on a standing army here within our own borders [...]. The world has changed. Today, our national security can be threatened in faraway places. It's up to all of us to be aware of the strategic importance of such places and to be able to identify them. (Reagan 1983)

Differently from the 1980s, the violent polarization of Lebanese society in the Two Years' War, however, was seen instead as a purely local issue that only necessitated enough amount of control to avoid it 'spilling' into regional escalation. Because the situation in Lebanon and its actors were not (or no longer) associated to any territorial danger to the international situation or loss of US political and ideological influence to the Soviet Union, then it meant that 'Lebanon was not perceived as being lost to the Soviet Union or to its clients; it was merely being lost to itself' (Salem 1992).

As seen in Chapter 3, critical geopolitics exposes the arbitrary workings of power that lay behind normalized geopolitical reasonings and representations of the world. In other words, it deconstructs how specific spaces are 'scripted' geopolitically. While geopolitical reasoning refers to the '*spatialization* of international politics [...] in foreign-policy practice' (Tuathail and Agnew 1992, 202), geopolitical scripts are defined as '[d]irections for the *performance* of global politics' (Sharp 1998, 159) [author's emphasis].

In 1975 and 1976, the representation of Lebanon in international foreign policy by international diplomacy can be explained as structured around three geopolitical scripts: the *tragedy* script; the *chaos* script; and the *fate* script. These normalized a depiction of the civil war as a tragic yet too complicated phenomenon, on a territory that had become 'lost to itself'. Its conflict was now represented as almost impossible to comprehend and it was therefore too complex to intervene directly in it.

## The tragedy script: Non-intervention as *territorial trap*

This first script implied, firstly, that the complexity and gravity of urban warfare in Beirut were overlooked by foreign policy practice in favour of the preservation of wider regional and international geostrategic equilibriums. Secondly, that the non-state armed militias fighting on the ground were rarely considered as significant actors with meaningful agency and power. Rather, the war in Lebanon was a *tragedy*, with no causal agency, and which only the Lebanese were responsible for tackling.

Agnew (1994a, 1995b) calls 'territorial trap' the combination of three geographical assumptions on which conventional international relations rely on to represent the space of global politics. These are: 'states as fixed units of sovereign space, the domestic/foreign polarity, and states as "containers" of societies' (Agnew 1994a, 54). The predominance of the territorial state is pervasive in geopolitics, where issues of security and defence are shaped almost exclusively along 'foreign' and 'domestic' affairs, making it almost impossible to elude from national borders when analysing world politics. This notion is based on classical definitions of sovereignty as the nation state's 'monopoly on the legitimate use of violence within its territory' (Biersteker and Weber 1996, 14) — and 'to enforce internal order and to protect against external threat' (Agnew 2009, 6). As it is broadly accepted in International Relations, sovereignty is considered an achievable condition of exclusive and homogeneous state authority over a territory. In this perspective, the nation state is the sole legitimate component, the 'universal unit of speech' of any discourse about and practice of global politics.

Consequently, the territorial trap outlook depicts any irregular non-state actors as anomalies to extirpate, therefore excluding them from legitimate configurations of world politics. Political geographers have often highlighted the inadequacy of the Westphalian (or state sovereignty) model in a globalizing world (Agnew 2009; Agnew and Corbridge 1995) and have argued that different actors and polities existing alongside the state need to be accounted for. Moreover, the state sovereignty model creates 'structures of legitimacy' (Gregory 2006, 100) by placing state actors (such as regular armies) in a realm of legality and 'legitimate war' and nonstate actors (like irregular armed groups) in that of illegality and, terror (Elden 2009).

Unsurprisingly, then, the geopolitical approach to Lebanon by several western governments between 1975 and 1976 was overwhelmingly state-centred, and the nation state acted as, borrowing Ó Tuathail's words, the 'omniscient narrator of territory' (Ó Tuathail 1996, 12). The characteristics of the irregular militias, as well as of the civilian population, rarely constituted the subject of international diplomatic talks and the specific urban dynamics through which these actors exercised power in Beirut were rarely acknowledged and often left vague, alluded to through mentions of 'fanaticism' and the 'irresponsible behaviour' of 'a relative handful of ruthless or hotheaded men' (British embassy Beirut report 1975).

And yet, a closer look at declassified diplomacy documents, such as the cables coming from the British embassy in Beirut, reveals that there was much detailed information coming from intelligence agents about the geography of the violence and division in the city:

> From 13 to 16 April the fighting was concentrated in the suburbs of Chiah, Ain Rummaneh, Sin El Fil and Dekwaneh, an arc reaching from near the Sabra/Chatila Palestinian camps south of Beirut to near the Tel Zaater camp north-east of the town. ('British Embassy report, 19 April,' 1975)

Particularly, intelligence accounts of specific violent incidents reveal the intentions of the fighters and, more importantly, the concrete impact that the conflict was having on the urban fabric in terms of economic targeting, and the developing confessional division:

> The bombings, which have concentrated remarkably on furniture shops and factories, were deliberately planned to hit one of the most important parts of the Lebanese export trade and so destroy the Lebanese economy and enterprise with it. This is plausible, but the targets might have been chosen simply because they were Christian owned and sold luxury goods. (British Embassy report, 5 July 1975)

One of the consequences of the territorial script in the evaluation of the Lebanese civil war by diplomats and foreign policy representatives, is the depiction of the 1975–1976 war as an exclusively internal matter, born in

Lebanon, developed in Lebanon and, therefore, the responsibility of Lebanon. Structural categories in international relations 'internal' and 'foreign' affairs are often insufficient to account for the multi-scalar and transnational nature of some of the actors involved in the Lebanese conflict. With the exception of Israel's then Minister of Defense Shimon Peres' address to the Knesset on 6 January 1975, which detailed the entity of 'Fatahland'[3] in south Lebanon, where Palestinian fighters had operated across the border with Israel, the Two Years' War was portrayed in international politics in a rather more simplistic way, as a local phenomenon. Even if the transnational connections of the militias operating 'both in Lebanon and outside' (Israel Ministry of Foreign Affairs 1975) and their being 'supported from outside' (White House 1976a) are mentioned in various occasions, the war itself is described as belonging to the realm of domestic matters. During an interview, then French Prime Minister Jean Sauvagnargues[4] noted that 'The problem in Lebanon is an internal problem, it's a problem of renovation of the Lebanese structures; therefore, it's a matter that concerns the Lebanese' (Ministère des Affaires étrangères 1976b).

And if the problem was internal, then it needed to be dealt with inside the territory of Lebanon. In other words, the accent on the nation state as the sole unit of interpretation for the Lebanon conflict, turned the preservation (or respect) of the territorial integrity and sovereignty of Lebanon into the utmost priority of the 'disinterested' (Ministère des Affaires étrangères 1976b) foreign policy agendas of the international community. State representatives were adamant that they could do nothing except state their opinion without intervening in what was seen as a conflict linked exclusively to Lebanon's domestic affairs:

> Concretely, we cannot send troops! Nobody is asking us to do that. Let's not forget that Lebanon is an independent State. [...] We cannot but contribute to create the favourable conditions for a political solution, so that a political compromise can be found. (Ministère des Affaires étrangères 1976b)
>
> The policy of France with regard to Lebanon is inspired by one only desire: that of preserving the unity, the integrity and the sovereignty of this friend country. (Sauvagnargues 1976)

A confidential cable sent by the UK foreign mission in Washington to the UK Prime Minister on 3 November 1975 about the situation in Lebanon, stated the US position with regard to intervention and invitation to other countries to do likewise.

> In briefing senators on 30 October [1975] Atherton (Assistant secretary, State Department) gave assurances that he US was not contemplating

intervention and was continuing to urge other countries to stay out of the conflict. The US attitude remained that the problem was an internal one in which western governments could do little more than urge those on each side of the conflict to be flexible. (Foreign and Commonwealth Office 1975b)

This is where the dominant geopolitical script towards Lebanon among Western diplomatic circles starts to look like a morality tale. With the Lebanese civil war depicted as a purely domestic matter and with foreign intervention ruled out as a matter of respect of Lebanon's sovereignty and territorial integrity, the international community could only express its sympathy and recognize the conflict as simply an unhappy turn of events – a tragedy. Moral judgements regarding the inhumanity of the situation in the country were frequent, but they also recognized the impossibility of the international community to act militarily because the events were regarded as exclusively Lebanon's internal affairs and mingling with them would mean to interfere with the sovereignty of an independent nation state. Diplomatic statements abound with allusions about the inevitability of this conflict and the impotence of external actors in changing a seemingly unstoppable course of events. Describing 'the agony of Lebanon', the most common expression was that of the country constituting a 'severe human tragedy' (Israel Ministry of Foreign Affairs 1976): 'Lebanon is a tragedy' (White House 1976b), declared US Secretary of State Henry Kissinger in a Cabinet meeting with President Ford on 18 June 1976.

In a rather different statement, Shimon Peres denounced, however, that the world was acknowledging this tragedy in 'apathetic and indifferent' ways, while Lebanon was 'wallowing in its own blood' (Israel Ministry of Foreign Affairs 1976). The moral rhetoric about the drama of Lebanon was never accompanied by any plan of intervention and was limited to expressions of 'amicable support' and 'deep sadness and [...] preoccupation' (Paul VI 1975b); 'sincere emotion towards a friend country,' 'moral contribution', 'political support' and 'efforts towards the re-establishment of the conditions for dialogue' (Ministère des Affaires Étrangères 1976c).

Vatican emissaries Cardinal Paolo Bertoli and Archbishop Mario Brini travelled to Beirut as peace envoys, in November 1975 and April 1976 respectively. The Vatican diplomatic endeavour in this phase of the war did engage with specific actors around issues of national sovereignty, however, the urban fighting and the division of the city were not tackled directly and the focus of the discussions was centred on the question of the national boundaries. Notably, dissuading the Maronite clergy and notables from advocating the sectarian partition of the country via the return to the *Petit Liban* (Mutasarrifiyya) and on responsibilizing neighbouring states to act to solve the Palestinian predicament (Frenza 2008). Sectarian coexistence was seen by the Bertoli mission as a 'universal model' (citied in Frenza 2008, 5), the original recipe for pacification, to which the Lebanese people had to return

Making the Lebanese population responsible for finding a solution to the war reflects the classic binary between domestic and foreign affairs in international foreign policy. As the problem was supposedly internal to the Lebanese territory and the conflict stemmed from its own society, it was somehow expected that the Lebanese people had the duty to find a way out of it. This point emerged in the statements of French Foreign Minister Sauvagnargues who, on various occasions argued that: 'it is up to the Lebanese themselves to find beyond every external interference a political solution that can put an end to the drama that divides them' (Sauvagnargues 1976) and also in the speeches of his successor De Guiringaud who, with a strikingly similar rhetoric, prompted that 'it is up […] to the Lebanese themselves to find within dialogue the political solution that can put an end to their fratricide struggles' (Ministère des Affaires étrangères 1976c).

According to this perspective, the Lebanese must renounce violence and 'must become conscious that they are *the only ones* capable of stopping the fatal mechanism that is conducting their country to the breakdown' (Ministère des Affaires étrangères 1976b, 1). Also, according to the Vatican, the Lebanese must 'renounce […] with generosity and perceptiveness, the fighting and the destruction' (Paul VI 1975b). In these views, the Lebanese are the only ones who can bring *their* nation back from the path of sectarian fighting. In other words, the situation was described as an irrational diversion that happened because of 'irresponsible' elements acting in the name of fanaticism leading the State to a breakdown. The Lebanese population, therefore, was called by the international community to return to a 'balance' in Lebanon's internal affairs, to quell the unrest and to devise to the solution of which they only had the key. The classical conception of the nation state as a territorial container of society emerges from these statements: the territorial nation state is a modern 'imagined community' (Anderson 1991) projected towards the same values and sovereign over a population contained on a contiguous territory. The call for the Lebanese to solve 'their own' drama is not only territorial, but also assumes a natural attachment of a people sharing the same values and the same land – an overly simplistic depiction of what was instead an already deeply polarized society as seen in Chapter 4. The belief in the moral and political amalgamation of society and territory is at the basis of the geopolitical representations of the

Lebanese conflict: the war is considered as a problem of reconciliation, one implying the need to make an effort to regain an original national unity that has been lost. Rather than seeing it as a problem with contextual causes and traits, international diplomacy, rather simplistically, scripted Lebanon's civil war as a morality tale of deviation from an original situation of coexistence.

## The chaos script: The politics of loss

The state-centred perspective, and the confinement of the war to a domestic matter internal to the sovereign state, and therefore not worthy of external intervention, also – and somehow paradoxically – contributed to shape a second geopolitical script. The 'chaos script' can be identified in the recurrent statements about Lebanon having become a 'lost' country, where a 'confusing', even 'illogical' conflict is puzzling the international community and preventing it from acting. This geopolitical imagination of Lebanon was that of a lost state, drowning in a self-made deadly tidal surge of violence that only the Lebanese knew how to stop. During 1975 and 1976, the Lebanese Armed Forces (LAF) were losing their monopoly of legitimate violence to a number of irregular armed militias fighting out their differing political and territorial agendas on the urban ground. In the geopolitical imagination of several western governments, Lebanon became then a country that could be no longer explained according to the territorial logic of the nation state. The breakdown of the nation state institutions dissolved therefore 'the necessary space of politics' (Minca and Bialasiewicz 2004, 92). And, in this view, once the space of politics was gone, what remained was simply chaos that was impossible for the international community to rationalize in the territorial terms that are necessary to the discourses of global politics.

Shimon Peres, for example, used a feral metaphor to explain Lebanon's loss of territorial sovereignty to armed irregular militias and descent into wild chaos, comparing Lebanon to a body ravaged by actors both illegitimate and irresponsible. The Lebanese nation, he declared, was *'being eaten away* by a collection of terrorist organizations devoid of any responsibility, and jeopardized by camouflaged intentions to *devour* her independence' (Israel Ministry of Foreign Affairs 1975). In Lebanon, reports an intelligence cable to the British Embassy in Beirut, 'it is now virtually impossible to say who is fighting whom' ('British Embassy report, 5 July,' 1975, 3). Chaos, uncertainty and difficulty to discern any future scenario seemed to reign in 'the complex Lebanese tangle' (Foreign and Commonwealth Office 1975a), as stated by Peres at the Knesset in 1976:

> Lebanon's more distant future is still shrouded in mist. No-one can foretell with certainty whether the human tragedy that has attended Lebanon has already in fact come to its end, or whether and how Lebanon will be reconstituted. [...] One thing, I believe, is meanwhile clear beyond doubt: the former Lebanon, as we have known it over the past 50 years, will be no more. (Israel Ministry of Foreign Affairs 1976)

The geopolitical discourses of international relations are not limited to constructing coherent visions of the world, but also impact on its very experience, entering 'directly into its fabrication' (Gregory 2004, 121) and are 'centrally involved in the actions of soldiers and cannons too' (Gregory 2006, 89). Likewise, the geopolitical script of Lebanon as a fragmenting country in chaos constituted the foundations and the justification of a policy of non-intervention right when the capital of Lebanon was becoming divided for years to come.

During the Two Years' War, then, it was generally agreed among the international community that the situation would not be solved by 'pushing troops into Lebanon' (White House 1976a, 2). The only interventions within Lebanese territory were the evacuation of the foreign embassy staff and the dispatching of various diplomatic envoys to Beirut. In this early phase of the war, the geopolitical significance of the conflict in Lebanon was almost exclusively related to the avoidance of a major Soviet involvement, as well as maintaining wider regional Arab-Israeli peace. This non-interventionist stance is glaringly apparent in the valedictory dispatch of British Ambassador to Lebanon Paul Wright just after the civil war started:

> Lebanon's present importance is not great, but a change to an anti-Western regime would harm our interests. We should help to maintain the status quo. (Wright 1975)

Whereas Lebanon's importance was reputed minor in the wider system of Middle Eastern international relations, domestically instead, Lebanon was depicted it as a country whose future was obfuscated by the mist of an intricate civil strife. The geopolitical representation of Lebanon as puzzling and complicated for global politics was, however, not new to US diplomacy, for example. Lebanon occupied a prominent position in US diplomatic concerns since 1958 (Salem 1992) after postcolonial rule elsewhere in the Middle East like Cairo and Baghdad were swept by Socialist regimes and American fears of anti-Western regional escalations started to increase. Clashes between progressive and conservative armed militants spread in 1958 in and around Beirut, leading to the US Marines' 'Operation Blue Bat' inside the city aimed at curbing the skirmishes (Chapter 3). Whereas on one hand the operation was clearly important for averting Lebanon's potential for regional instability, on the other hand the organization, actions and purposes of the militias on the domestic ground remained something perplexing for the US government. This involvement revealed to US diplomats and policy-makers the complexities and contradictions of Lebanese politics of which, until then, they had known little and vaguely. The Lebanon perceived by the United States after 1958 was one of bewildering diversity of faiths and shifting alliances (Salem 1992). In the geopolitical imagination of the mid-1970s Lebanon therefore remained an entity 'so complex that it defies logic' (White House 1976a, 2). The following conversation between US President Ford, then Secretary of Defence Rumsfeld and National Security Advisor Scowcroft about a congressional briefing on Lebanon illustrates this challenge to grasp the situation on the ground:

Rumsfeld: The situation particularly defies logic as Henry [Kissinger] tried to explain it to the Congressional leadership this morning.
President: They came in confused about the situation, as they usually do.
Scowcroft: And they left still confused but at a higher level.
(White House 1976b, 2)

These references to the intricacy of the Lebanese conflict are more than neutral descriptions. They are 'form[s] of classification and particularization' (Ó Tuathail 2002, 615). Their use and repetition – like in a script – created a moral ground and a coherent discourse that justified non-intervention right at the time when the fighting was gradually partitioning Beirut, from its suburbs to the city centre and changing its demographics inexorably. This kind of categorization of a conflict as 'complicated' is very similar to the Bush administration's strategy towards the Bosnian war, which was condemned in the *New York Times* as 'hiding behind a self-perpetuated veil of confusion and complexity' (Ó Tuathail 2002, 615).

The practice of foreign policy towards Lebanon at the time was, unsurprisingly, caught in the territorial trap by which 'in idealizing the territorial state we cannot see a world in which its role and meaning change' (Agnew 1994a, 77). Once the Lebanese territorial state crumbled under the weight of the irregular militias, what remained was a logic-defying country and its domestic war, represented through images of chaos. Lebanon became a space where international relations could not perform in territorial terms and where territorial solutions failed because sovereignty was lost. The state here was no longer the sole and legitimate narrator of Lebanese politics, let alone the actor in control of political violence. Official foreign policy did not even attempt to consider the geographies and territorialities of the militias, and confined them to a murky realm of undisciplined irrationality, as 'irresponsible' and 'extremist' (Israeli Minister of Foreign Affairs 1975).

In a conversation with the members of the National Security Council, Henry Kissinger stated the attempt to create a geographical alternative, which however was aborted because, again, of the lack of a legitimate political interlocutor in Lebanon:

> we have explored the idea of a neutral zone, but there are too many undisciplined, criminal elements and there is no one to police them. A buffer zone without force is no good. (White House 1976a, 9)

In a call for international diplomacy to turn its gaze upon Lebanon, French officials exposed the worsening spiral of darkness and chaos in which Lebanon was kept by the warring factions:

> the information that we receive is confused and contradictory. Only one thing is sure: that there are more and more victims among the civilian population and that this unfortunate country is more wounded everyday, and that all the concerned states will have to eventually comprehend that they will have to think about the safety of Lebanon and that this will be in everybody's interest. (Ministère des Affaires étrangères 1976a)

The metaphor of the Lebanese nation as a wounded body returns in the French description of the aggravating bleeding of Lebanon and the death of its population, a condition which should push the international scene to act in order to re-establish the 'safety' of the nation state, for the sake of 'everybody's interest', but especially avoiding a major regional war.

## The fate script: The unavoidable sect

In Chapter 2, we saw how modern territorial governmentality contributed not only to shape the present-day boundaries of Lebanon, but also to establish the idea of the sect (Taifa) as a state-recognized component of the institutional power mechanism of the modern nation state since 1936. This vision became dominant, and political relations based on confessional belonging overtook feudal power mechanisms and hierarchies previously mostly structured along lineages of blood and family (Makdisi 2000), but that soon lost their contextual (and specifically colonial) connotations and became instead normalized and crystallized as an age-old second nature of the Lebanese (Makdisi 2000), unavoidable as dictated by birth.

This provides the historical context for the third geopolitical script that resonates in official foreign policy practice towards Lebanon in 1975–1976. It is centred around statements about fate and the inescapability of the sectarian nature of Lebanon, recurrently used by official foreign policy representatives to make sense of the situation and, ultimately, it reinforces even further the policy of non-intervention. In a sort of sectarian determinism, the confessional community replaced the nation state and became the new – fatal and inescapable – unit of speech to narrate the 'chaotic' Lebanese events, where political leaders seem to act almost exclusively out of 'sectarian nature' rather than according to specific and contextual strategies of power.

One recurring point in the fate script is its view of Lebanon as a country populated by inherently different, but coexisting, sects. Because these differences took over coexistence, the country, however, plunged into discordance and chaos. According to these scripts, these innate differences between sects form the fundamental unity of Lebanon, a unity constituted of the sum of the numerous, diverse and discrete *communautés* populating the area through its history, and of 'all those who, with their diversity, compose this country' (Sauvagnargues 1976). The country, defined as 'the sole multi-communal state that existed in the Arab world' (Israel Ministry of Foreign Affairs 1976), was represented as a lost paradigm for multicultural harmony, especially by French officials, whose predecessors helped to create Lebanon as a modern state – and to officialize and make those communautés the main components of the mechanisms of government. They depicted Lebanon as 'a small model-state, an example of tolerance in the Middle East, a state that we largely contributed to create' (Ministère des Affaires étrangères 1976b, 1) and 'formerly a model of coexistence between different communities, left for little more than a year to an uncontrolled outburst of violence' (Ministère des Affaires étrangères 1976c).

The religious sect as the basis of national governance and territorial unity was constantly recalled in the geopolitical statements about the civil war. 'The Christian community in Lebanon' was depicted as 'Confronted by the Moslem [sic] community' (Israel Ministry of Foreign Affairs 1976), and the sectarian balance between the communities 'surely different in their origin and features, but united in an intense activity for the love of the homeland' (Paul VI 1975b) was depicted as the condition to which Lebanon should return in order to be granted peace and national reconciliation in that 'exemplary stage of fraternity and collaboration' (Paul VI 1975b) that characterized it. In particular, the Christian confessions were framed as age-old, bounded and distinct communities by Peres who described them as 'implanted in its soil for countless generations' (Israel Ministry of Foreign Affairs 1976) and by the Vatican by emphasizing 'the century-long ties uniting Lebanon to the church, through a history that goes back until the Apostle Saint Paul' (Paul VI 1975a).

The agency of the Lebanese militia chiefs and political personalities has also been conceived as solely determined by their belonging to a sect. Their actions are seen as exclusively inspired by sectarian forces, and void of any socio-spatial context. Henry Kissinger's explanation of the Lebanese scenario describes the National Movement (and Druze) leader Kamal Joumblatt as a person whose 'natural inclination will be to destroy the Christians' (White House 1976a, 3). This is a primordialist vision of sectarian violence, which assumes the a priori existence of the sect outside the social process and considering the events of the war as pure consequences of sectarian belonging. Here, 'the resilience and tenacity of Lebanon's primordialism' (Khalaf 2002b, 210) is reinforced by the international community's geopolitical scripts to justify a policy non-intervention.

Deterministic tropes about unstoppable chains of feudal violence and 'natural inclinations' to kill conceal the complexity of the events on the ground. By the end of 1975, the war was assuming the dimensions of humanitarian disaster: street fighting, snipers' fire, random killings, sieges of neighbourhoods and bulldozing of refugee camps (Yassin 2010), kidnappings and summary executions. Instead of analysing its contextual and urban traits, official foreign policy representatives explained it away in mechanicistic, disembodied and ultimately vague terms, such as 'implacable mechanism' (Ministère des Affaires étrangères 1976c), 'fate' (Israel Ministry of Foreign Affairs 1976), or even a 'fatal gear dragging the country to explosion' (Ministère des Affaires étrangères 1976b, 1). What was re-formulated in official foreign policy as the irresponsible actions of a bunch of irrational thugs, constituted instead not only wanton, but very often thoroughly planned violence against the spaces of the civilians' everyday life and homes, as documented by the intelligence reports from the ground operated both by the Kataëb militias and by the rival progressive ones:

> At least two similar actions, in which the Kataëb appears to have taken part, took place over the last 10 days. The pattern is one of relatively large scale attack on a[n] isolated locality, inflicting some, but not usually many casualties, the destruction of individual houses by burning or bombing, and the terrorisation of the inhabitants so that they are induced to leave. Between 400 and 600 inhabitants of Sibniye fled and are now thought to be housed in beach chalets on the southern outskirts of Beirut ...

Progressive forces took retaliatory action yesterday and claim to have blown up 33 houses in Kfarshima.' (Foreign Mission in Beirut 1975)

These representations of Lebanon's space silenced other geopolitical dynamics that were changing the region and that shifted international foreign policy away from interventionism, such as the 1973 oil crisis, the increasing power of Saudi Arabia and Iran, the priority of Israel–Palestine peace process, and the reluctance towards military intervention after the Vietnam experience. More importantly, these depictions of Lebanon took no interest in the urban actors and spatial changes that shaped the future of the country's capital for years. The analysis in this chapter has highlighted an *urban lacuna* in the practice of foreign policy towards Lebanon: a lack of appreciation of alarming events unfolding at the urban scale, in favour of a state-centred view preoccupied with the maintenance of the regional status quo and of primordialist visions of sectarian belonging as the only factor of militia violence. Armed fighting, the dismemberment of the army, the takeover of key buildings (as we will see in the next chapter), population displacement into homogeneous territories, and eventually the partition of the city, were concealed behind the aura of primordialism and inevitability with which the unfolding of these events was depicted by international foreign policy. While vast and detailed urban local knowledge of events, actors, and places featured at length in intelligence cables at least from the British agents, this did not translate, in the upper diplomatic circles, into attention towards the city fragmenting, let alone into action to prevent it from dividing. The massive changes in Beirut's urban territoriality did not enter official foreign policy reasoning, but did, however, crystallize social polarization into a divided urban fabric for the rest of the war and beyond.

Beirut was changing inexorably: while several spaces were targeted by wanton violence and disappeared, new spaces emerged – dividing lines guarded by snipers, physical barriers, checkpoints, no-man's lands – that now punctuated the fragmented everyday landscape of those inhabitants who still lived amidst the by now protracted conflict. Elsewhere, however, in the circles of official foreign policy, a different script was now established: one where Lebanon as a nation state had been eclipsed. What replaced it was simply nothing more than a tragic but unstoppable and incomprehensible concatenation of violence by warlords driven by sectarian inclinations.[5]

Urban reasoning and vocabulary were absent from these scripts, in favour of state-centric ones, specifically preoccupied with containing foreign state actors from entering Lebanon and from upsetting the balance of power between the Western and Soviet blocs – despite the fact that armed nonstate actors were already receiving foreign funds and recruiting members from other Arab countries (El-Khazen 2000, 302–03 Table 22.4. Source: Lebanese Army Intelligence Report, 17 July 1975). The war was also scripted using a narrative of senselessness, and even deemed 'illogical', despite detailed intelligence daily reports from the ground about the reasonings of armed militias, and of how, where and by whom violence was being perpetrated. Finally, sectarian radicalism was singled out as the only determining factors in the spreading of violence.

These geopolitical scripts and their capability of shaping policies of non-intervention by Western governments, in contrast to Operation Blue Bat in 1958, were a powerful but partial and selective representation of the war. There were, however, other geopolitical visions, that were taking shape in and through the brick and mortar of the buildings, streets and homes of the urban battleground that Beirut had become. It is on these other scripts, and the urban sites in and through which they acquired shape, that the next chapter focuses. Chapter 6 takes the analysis into the discourses and practices of urban warfare that official diplomacy statements silenced or summarized as unintelligible chaos or sectarian mechanism. Here, Beirut is no longer a passive backdrop for proxy wars or incomprehensible fanaticism, but the very site where other geopolitics, differential ideas of the Lebanese nation state and its relation to the world were shaped and produced through the re-drawing of new urban territorialities.

During the Two Years' War, the monopoly of the Lebanese state on political violence and its own territory became increasingly feeble, with the fragmentation of the Lebanese Security Forces, and the establishment non-state actors with power and control of specific urban territorialities. It is in and through these new territorialities, and through the reworking of the physical urban fabric through violence on the urban battleground, that new geopolitical meanings were produced about Lebanon's presence and role in the region. It is to these urban geopolitics that I will now turn.

# 6

# De-Subjugating Beirut's Urban Geopolitical Knowledges

*Now I understand my father, who would remain at home smoking all day long and drinking coffee, waiting to see me return. (Edouard).*

*I would like to point out that, in my nature, I don't like violence [...] Sometimes circumstances impose themselves on you [...] sometimes one feels that it is his [sic] duty to take responsibility. (Nizar)*

The previous chapter offered a genealogical account of the geopolitical scripts that underpinned non-intervention in Lebanon in 1975 and 1976. At the time when statements shaping those scripts were released by foreign policy officials of several countries including the UK and the United States, the division of Beirut was under way, and wanton violence was redrawing its urban political geographies. By contrast, the diplomatic statements and international official addresses about the Two Years' War lacked consideration of the urban dynamics and implications of the conflict. These binary and reductive, yet powerful, geopolitical scripts hindered intervention in what was portrayed essentially as a quagmire and allowed the Lebanese capital to become physically divided according to the rationales of nonstate armed groups. These rationales, and the spaces they produced, constitute the focus of this chapter, which offers a contrapuntal reading (Said 1994) of the geopolitical scripts in the previous chapter and emphasizes the differential accounts of state, nation and power shaped amidst the urban battleground in Beirut in 1975 and 1976.

As French geographer Yves Lacoste (1982) argued, the geopolitics of statecraft is not the only legitimate one; there are also 'other' geopolitics, through which the space of international politics can be narrated, mainstream geopolitical discourses resisted, and the contested and partial character of official geopolitics exposed: 'Geopolitics is not only about planetary-wide considerations about the strategies of the superpowers: it's also the reasoning that can help to resist to their hegemony' (Lacoste 1982).[1] In Beirut, one of many cities in the world which have endured militia warfare, urban space was not just a backdrop to the sectarian battles between 'ruthless or hotheaded men' (British embassy Beirut report 1975) who incomprehensibly deviated from the social pact. *Other geopolitics*, differential imaginations about the Lebanese nation state and its relation to other states took shape in Beirut during the Two Years' War, and urban space was a key agent in their formation. Other, subjugated geopolitics, moulded out of the micro-geographies, the micro-universe of words, practices and materialities of nonstate actors fighting on the ground, borrowing from and speaking back to the macro-worlds of official geopolitics.

The following pages give voice to and map some of these subjugated knowledges (Foucault 2003), differential geopolitical accounts about Lebanon and Beirut in the map of global politics, that are not part of the official geopolitical accounts of the war. They bring frontstage the constellation of subjugated – and highly contested – geopolitical imaginations, that is born not from the classic geopolitical considerations about inter-state relations and balance of power typical of the Cold War geopolitical episteme (Moisio 2015) seen in the previous chapter, but out of the reasonings and practices of nonstate actors engaged in urban warfare. These subjugated constellations of 'local, regional, or differential, incapable of unanimity' (Foucault 2003, 7–8) geopolitical knowledges are unpacked and exposed here appreciating how mechanisms of semiotic representation and human-based production of meaning (e.g. militia statements, slogans, visual practices and visual propaganda) are intertwined with non-human registers of conflict (objects, maps, vehicles, ammunitions and of course the physical materialities surrounding: high towers, check points, and so on). This way, the chapter aims to open up the perspective onto a non-anthropocentric (Coward 2006) understanding of violence and thus ground subjugated geopolitical knowledges within specific urban spatialities.

## The built environment and the propagation of violence

At the basis of Martin Coward's work on the 'politics of urbicide' (2009) during the Balkans war is the idea that violence against the built environment does not accomplish itself just in destruction, but rather, in the consequent obliteration of what Coward – drawing on Heidegger's notion of *Dasein* – sees as the essential condition of urbanity: heterogeneity. Since buildings, argues Coward, 'constitute the spatiality of existence as fundamentally shared/public […] destroying buildings is essentially a destruction of the conditions of possibility of heterogeneity' (Coward 2009, 72). Omar's views about the reasons for targeting the buildings in the city centre of Beirut echoes Coward's conceptualization:

Beirut's city centre was a place of mix for the entire Lebanese population, based on their shared interested. The destruction was carried out in the centre of Beirut in order to eliminate the meeting point for the Lebanese population, so that they wouldn't meet based on mutual benefits and interests, and so that everyone would go back to their sect and religious group. It was carried out to promote a division in the Lebanese population at the sectarian level. (Omar)

In his analysis of the constitution of social relations in cities, Isin (2002) describes how, far from being disembodied communities of identity, 'groups cannot materialize themselves as real without realising themselves in space, without creating configurations of buildings, patterns, and arrangements, and symbolic representations of these arrangements' (Isin 2002, 42–43). In other words, 'space becomes its defining characteristic as a group' (2002, 44), part of the mechanisms of collective life and therefore 'is a condition of being political' (2002, 43). In this chapter, I disentangle the attacks against the built environment in Beirut from pure paramilitary strategic considerations (such as the presence of rival militias, their position vis-a-vis the dividing line) and from fixed symbolisms (such as the meaning of a building for a certain sect). I treat instead the attacks against the built environment as a means to reinscribe profoundly the relations between people and things in the city: 'We rebuild, it's true. But it's no longer as it was before. It cannot any longer be like before', said Edouard towards the end of our conversation, reflecting on the long-term impact of the violence on the city. In other words, the attack against the built environment is a means to redraw long-term the city's spatiality, the very specific relations between people and the things that surround them, and ultimately against the city as a machine of differentiation and identification (Isin 2000).

The successive rounds of militia warfare during the Two Years' War contributed to the demarcation lines becoming fixed 'and they did not shift anymore' (Corm 2005, 212). This led to Beirut's partition into two sectors whose demography was dictated by sectarian belonging. As Omar again explains, 'the occupation of buildings was functional to the phase between the start and end of the battle of the hotels [October 1975–March 1976]. This was a war for the occupation of buildings and the advancement of the frontline. After this, it became a war of fixed positions".

As the war continued, urban space became the materialization of contrasting geopolitical visions, but it was also part of everyday paramilitary tactical considerations, as its very presence influenced militia warfare. Moreover, paramilitary violence affected not only governmentally and/or religiously symbolic buildings and spaces, but also mundane ones. The politics of urbicide rely on targeting the built fabric that mediates everyday social relations, and not only on the destruction of a community's symbolic buildings. More specifically, urbicide is the attempt to eliminate the very spatial basis for heterogeneity and for the coexistence of difference (Coward 2009). The coexistence of tactical consideration of urban space, with its symbolic and political aspects reoccurs in the accounts of the fighting and in media chronicles of the time, often without a clear-cut distinction between what was perceived as destruction as the collateral damage of tactical needs and what was instead considered as destruction as a (geo)politico-territorial statement.

The geopolitical discourses underpinning the fighting, rather than being an all-encompassing driving force, developed in parallel with the way fighting progressed in light of tactical and even everyday pragmatic considerations. As specific neighbourhoods became increasingly contested and the necessity of maintaining territorial control increased, the fighters' own involvement in political causes extended and the intensification of military activity in the neighbourhoods developed in parallel with the increasing politicization of ordinary residents:

> I was a resident of al Shiyyah, and the main axis of the fighting was around al Shiyyah. In the beginning, I was not really speaking politics. But then I got trained [in the use of] weapons, and I was seeing people discussing national issues, social issues, and things like that. In the first period, there was a faction – the Kata'ib – they would carry out attacks against [our] area, and the other faction here used to counter attack. The least I could do was to protect my house, so I trained, and then I started speaking politics. (Nizar)

The reconfiguration of urban territory in 1975–1976 through the creation and stabilization of the main confrontation line (Green Line), involved targeting high-profile and/or strongly symbolic buildings (such as the Parliament and big company HQs), but also informal ones (such as shops):

> On Paris Avenue, five passengers in a car rolling at high speed opened fire with their machine guns against an automobile exhibition hall belonging to Michel Mitri Anid. Numerous among the exhibited cars have been hit by the bullets. [...] The fire has reduced into ruins the Librairie du Liban and the headquarters of the Al Itimad bank, [...] as well as a part of the PanAm building. (L'Orient le Jour 1975b)

Very often, the presence of the Green Line was stated not by physical divisions, but by the snipers' action: the border between different urban zones was marked by maintaining the space empty by killing everything that moved. What in peacetime were urban infrastructures of connection like the Ring – the east-west flyover passing where the north-south green line met the city centre – became deserted no man's lands, where cars and people passed at their own risk. A former fighter aligned with the Kataeb militia recalled how

> the snipers were shooting on the Ring without knowing the religion of those passing by, of their victims. They would shoot on everything that moves […] Nothing could move, even the dogs. I have met people who describe how they took big pleasure in seeing a person fall dead within a matter of a second. This was the Ring. (Abu Layla)

Army detachments were deployed to protect official landmarks such as the parliament building and the annexed square, Place de l'Etoile, but eventually the militias entered and looted the Parliament and, in March 1976, the Cabinet building (*Grand Sérail*) was also shelled and occupied as part of the fighting for the control of the city centre for the stabilization of the frontline in early 1976:

> We carried out raids against the Sérail, a Phalanges spokesman specified, who has indicated that these forces did not occupy the block of buildings on a permanent basis. (L'Orient le Jour 1976)

Contrary to the representation of the civil war as a confusing quagmire and of the militias as fanatic, irresponsible agents as we saw in Chapter 4, the materialities of the fighting are the product of specific logics. Visuality in particular was an important aspect for the tactical reasoning of the militias, such as domination of the high-rise hotels on the seafront. Altitude and the capacity to control urban space from above, was certainly an asset in the battle for the 24-storey Holiday Inn hotel building, from which snipers and fighters from both militia fronts held their position in the battle for the completion of the green line:

> [The militias] believed [the hotels] were strategic buildings where they can use their height. At the very beginning, I remember quite well, when the gunmen came into the Holiday Inn, […] they asked me to open the elevator for them to go on the roof and shoot the other side from the roof. (Holiday Inn former engineer, June 2011)

Altitude provided the privileged point of view to grasp the city space and transform its complex social and spatial fabric into a tactical terrain to dominate. Contemplating and controlling Beirut from the height of modern towers, looking down at the city in the process of being divided in sectors in the name of an orderly architecture of enmity – a representation made real – the gaze of the militiaman is not only tactical, but also taxonomic. It reifies the architecture of enmity based on differing geopolitical imaginations about sovereignty and the borders of Lebanon, as well as the sectarian traits of these imaginations: East and West Beirut become a fact on the ground.

It is a well-established argument that colonial power sows its own resistance from within 'the organizational terrain of the colonial state, rather than in some wholly exterior social space' (Mitchell 1988a, xi). Anti-colonial movements are often born inside the institutions of the colonial state, such as schools and army barracks and use colonial modern devices (such as military training and modern education) to prepare, declare and often establish their presence (Mitchell 1988: xi). The mingling between militias and state actors, as seen in Chapter 4, included training with weapons by the Tanzim for the Lebanese Front side. In the first stages of war, as Rashid tells, the militias did not have salaries, were self-financed according to their own means and relied on charitable donations of food, drinks and equipment to sleep on the frontlines. Gradually, the militias started to use resources from institutions on which the State had no longer the monopoly:

> We had some individuals [...] associated with us, who were present in the ministries. I remember we had a comrade in the ministry of public works [...]. So [we used] as much as we could take from governmental institutions through people that we are associated with. (Rashid)

As the rounds of fighting continued, the militias increasingly used state resources to aid their movements and operations, as Nizar recalls in the case of maps ad movement between the urban frontlines:

> In the first phase, we did not have maps with us. It was done in a primitive way. Initially, there is action and reaction, they used to shoot from the East and we used to shoot back [...] But when the war later spread, there began to be movements (tana``ulāt), we started further battles [...]. Even the movements were primitive: I remember that the militia I was in did not have the logistics or the equipment. I remember moving in private cars, whereas in areas like al-Shiyyah we moved around the normal way, that is to say walking. Regarding the maps: now, that came in the later phases. The war developed and we started using maps. For example, in the two years war, I was involved ten [para]military rounds (dawra 'askariyya). In the first phases it was through primitive means: dismantling guns and stuff like that; then we were trained as lieutenants;

in the later phases of the war, around a year later, we started using maps and other engineering equipment. (Nizar)

As the war continued, the militias re-drew the space of Beirut using technologies proper to a state government: they collected taxes, structured their disciplinary system (including prisons), policing neighbourhoods, recruitment and basic urban services like refuse collection (Harik 1994):

> I quit my business, and I was completely dedicated to the military activities. […] In the first six or seven months we were paying monthly membership, to provide for ourselves and for the groups. We used to go around all the guys to make sure that they pay. Then […] there began to be financing for us: the resistance supported us, the Arab countries supported us and everybody financed, and then came something called Haraka al Wataniyya, there was support for the Haraka al Wataniyya so […] I started to be paid an allowance. (Rashid)

The geography of the city, and indeed its everyday built environment were rewritten in the language of militia warfare. Omar argues that combat transforms a building from a place of beauty to an object for a strategic mission: its function, its beauty are gone and it becomes a target. In much the same way, Beirut is turned into a geometry of military sectors, confrontation lines and axes: '[The militia chiefs] would request us as help in the battle if a particular axis was being pressured, for 2 or 3 or 5 days for an axis that is being pressured, and then I would return to the area where I was responsible' (Rashid). In this situation, not only the peacetime functions of buildings were re-programmed for war and the peacetime distinctions between domestic and public space, but the space of the familiar and the home and that of the battle and the city, became blurred. Elements of the uncanny and of danger spilled from the battleground and into the space of the neighbourhood, as in Abu Layla's account of his return to his neighbourhood of Gemmayzeh, what he calls *le quartier peuplé*, the populated neighbourhood:

> When we left the populated neighbourhood, we became the lords of the place. Alone, it's only us. It was another world, totally different, as if you were on the moon. When I returned to the populated neighbourhood, what was impressing me the most was seeing people live as usual: they laughed, they ate, had a very ordinary routine, the children go to their schools, the men go to work. But me, I came from the land of death, I have seen the other world, a crazy world. I almost died, I lost friends, I have seen people who are as young as me like me die. It was shocking to see. (Abu Layla)

Vice versa, the home would become a component in the geography of the fighting, in this case, as a base for recovering before going back to the fighting:

> When I was injured they would take me to Barbir Hospital. They would stitch my wounds, put a plaster, inject me, and I would return to my position. But I would pass by home to change my clothes, because [at this stage] we wore civilian clothes [and] they had blood all over. (Rashid)

Ultimately, the war reified the architectures of enmity that had been developing throughout the escalation phase illustrated in Chapter 4. The representations of difference as a danger, not only underpinned the violent actions against the city and its built environment, but re-envisioned spaces with complex socio-political genealogies into coherent and fixed political geographies of threat, hostility and enmity. This is the case with the spread of representations of the Palestinian camps among certain sectors of the Christian and/or right-wing population, according to whom the camps were surrounding and strangling Christian-majority areas of Beirut. As Edouard explained, 'they [the Palestinians] built camps around Beirut and more precisely around East Beirut'. While 'East Beirut' is a territorial concept born out of the civil war, it is reified here as something that precedes the war, rather than being produced by it, as if there was an East Beirut pre-existent to the war and to the creation of the Palestinian camps. The same mechanism was by no means present only on one side of the opposing parts, as retaliatory purges took place in both Christian- and Muslim-majority localities around Beirut. Locations subjected to this kind of retaliatory attacks were the predominantly Sunni Muslim Karantina (18 January 1975) and Tall al zaatar (August 1976); and the predominantly Christian Damour (20 January 1075) and Ayshiyeh (October 1976).

## Reframing sovereignty in the city

In Chapter 5 we saw how official geopolitical discourses replaced the complexity of urban warfare with a preoccupation for maintaining regional geostrategic balance in line with the Cold War doctrine of containment of both Soviet and of pan-Arabist influences on Lebanon. The armed militias fighting on the ground were often dismissed as extremists, deviating from the national pact of coexistence and dragging Lebanon into a messy domestic *tragedy* for which the international community had no solution. Abiding to the assumptions of what Agnew has termed the territorial trap (Agnew 1994b), the Lebanese tragedy was framed by the actors of international foreign policy as outside 'the necessary space of politics' (Minca and Bialasievicz 2004, 92) that is state sovereignty.

However, differently from the assumptions of these official geopolitical framings, when seen contrapuntally and from the urban level, the dynamics of violence of the Two Years' War blur the boundary between de-jure state sovereignty and de-facto sovereignty practices and arrangements of armed nonstate actors. The militias did not act in isolation from state power; quite the contrary, they spoke back to a gradually dissipating central state; they collaborated with some of its actors; and operated within specific wider geopolitical discourses, not denying them altogether, and dissipating them into chaos, but rather renegotiating them on the urban ground. These dynamics recasted the normative practice of sovereignty in the city, inexorably changing its spatiality.

The following passage is taken from a 1976 editorial for the magazine *Al Hawadess*, about the polarization and hostility developing between the two neighbourhoods of al Shiyyah and Ayn al Rummana in the south-eastern suburbs of Beirut. It reproduces the conversation between two rival militia members, explaining 'the Lebanese crisis' using two diametrically different geopolitical narratives – the pan-Arabist and the Lebanese nationalist:

> The foundations of the Lebanese crisis [lies in the fact] that the Muslims are not convinced that Lebanon is their ultimate homeland, that it deserves to be defended by us; and that its internal affairs are above any other of the other Arab issues. (Al-Lawzy 1975, 6)

These two geopolitical imaginations – Lebanon as a sovereign, bounded state on one hand, and Lebanon as part of a wider, Arab hinterland on the other, are key to de-subjugate the geopolitical knowledges produced during what at this stage was a predominantly urban conflict. Far from being an irrational quagmire, in Beirut specific architectures of enmity built around differing ideas of nation, sovereignty, and territory were being enacted at the level of the urban neighbourhood, through the reconfiguration of the urban built environment (via partition, roadblocks, sniping, demographic redistribution, and destruction) and the attribution of new meanings to it. The notoriety of militia leaders was directly connected to the neighbourhoods where their supporters were enrolled; these neighbourhoods, in turn, became one with associated ideologies, Murabitun leader Ibrahim Koleilat, for example, was defined as 'the new Mohammedan [sic] leader whose star is beginning to shine in the sky of Basta, of Mazraa, and even of Mousseytbe'[2] (L'Orient le Jour 1975a).

Well-established geopolitical notions about the sovereignty of the nation state, like the presumption of its monopoly of legitimate violence a within bounded territory, was not lost on or totally negated by the militias. In the early phases of the war, as we saw, the state still constituted a territorial and political reference: it was even invoked by some of the militias. After the mass killings of 'Black Saturday' in the centre of Beirut on 6 December 1975, for example, the leader of the Murabitun militia Ibrahim Koleilat called for strong government and army action, in the areas of Beirut where it still detained control: '[Koleilat] is astonished that the reaction of the state has not taken place in the moment in which the killers and the snipers raged within the same zones that had been invested by the army commandos' (*L'Orient le Jour* 1975a, 4). Koleilat also accused the state of mingling with the so-called 'isolationist' militias of the Lebanese Front, in connection with the aggression against the Saint-Georges hotel – the first building of the city's international hotel district to be occupied, when the National Movement took it over in late 1975. The army – according to Koleilat – had deliberately not prevented the Lebanese Front (whom he labels as 'fanatics') from attacking the building, where members of the Murabitun were barricaded after attempting to seize the rival militias in the close surroundings:

> [The Isolationists] attacked the St. Georges with incendiary bombs in order to displace the progressive elements. [The army] brought help to the fanatics when instead we had surrounded them in the sector Starco-Hilton-Zeitoun. (L'Orient le Jour 1975a, 4)

Inside the Lebanese Front, the role of the state also continued to be relevant and, instead of disappearing, its disintegration was exploited, and its functions reframed in favour the tactical needs of the militia. Part of the army – the implementing agent of the state's monopoly of political violence – gradually reorganized itself, to work alongside nonstate armed groups: the *Tanzīm*, as we saw in Chapter 4, was composed of army members training groups of young militia members to combat in the first half of 1970s (Farid El-Khazen 2000; Hanf 2015).

El Khazen (2000, 299) illustrates that the seven main militias of the National Movement had about 19,200 recruits, while the army had just 19,000. In early 1976, a rebellion at the army barracks in Al-Fayadiyya demanded that the President resign. This led to the army's polarization and eventual fragmentation. By then, training camps involving all Lebanese and Palestinian parties (El Khazen 2000) had developed in the mountains around Beirut and inside the camps, as Rashid recounts to me:

> I lived the situation before the war began, and even then anyone, even if they did not know about politics, were saying that this country is going to war, because the militias from all side were training ... we knew that

[the militias] had weapons and ... they were doing training courses (*dawrah*). (Rashid)

The militias were gradually negotiating a new relationship with state sovereignty. This new relationship was often shaped through representations of the city, its spaces and its built fabrics, and these representations were often supported and circulated via the material medium of the political poster (Maasri 2008).

Political posters are artefacts reproducing a certain experience of Lebanon, Beirut, the war, and its protagonists. They also produce meaning about the nation states and political entities beyond the boundaries of Lebanon, such as Syria and the Palestinian territories, Israel and the United States. Their visuality reproduces a different perspective on world politics from that of formal geopolitics and its technological abundance in military maps and other devices. Their perspectives expose the unevenness in the availability of visual technology (Haraway 1991) lying behind the scientific aura of geopolitical thinking and military cartographies. Posters entered the production of the social and material fabric of wartime Beirut. However, the 'scopic regime' (Rose 2001, 6) of the posters is also partial. It is embedded in the social practices of the conflict and their visuality is part of these processes. For example, the accuracy of their design, and the time frame of their appearance, could be linked to the increasing organization, structuring and economic stabilisation of the militias. For this aspect, I build on Chaktoura's (2005) study of mural graffiti, and particularly to her observations on the use of stencils rather than free drawing, the design of more refined fonts, and the repetition of phrases and slogans as the material translation of the passage from irregular rounds to continuous war, of the stabilization of the front lines, and of the establishment of the militias' fixed positions on the ground. Therefore, looking at the geopolitical visions that these visual sources convey, and highlighting the social processes in which they are embedded, is also an element for counteracting the scientific claim of a geopolitical discourse that overlooks the complexities of these sites of production of meaning during the conflict.

The poster in Figure 6 represents well the recasting of views of state sovereignty by the militias in relation to the conflict and its urban geographies. Although the production date and commissioning party are not given, according to Maasri (2008, 103) it can be deduced that it was published by militias belonging to the Lebanese Front at the time of the deployment in Lebanon of the Arab Deterrent Force (ADF, see chapter 5) in 1978/1979 the neighbourhood of Ayn al Rummana, depicted in the poster, came under heavy shelling in confrontations between the Lebanese Front

**Figure 6** Lebanese Forces propaganda poster. American University of Beirut, Archives & Special Collections. Political Posters. Poster No. 354-PCD2709–06.

and the Syrian troops, who had de facto become the only contingent of the ADF in Lebanon. Ayn al Rummana, however, was also a key landmark in the political map of the Two Years' War. As we saw in Chapter 4, the civil war is pinpointed as having started in this neighbourhood on 13 April 1975, and we saw in the account by *Al-Hawadess*, hostility between the neighbourhood of Ayn al Rummana and that of Chiyyah ran high, with Ayn al Rummana gradually becoming key territory of Lebanese Front militias.

The Arabic text at the base of this poster reads *The world is asleep while Ayn al Rummana stays awake.* The meaning of this line is complemented by the rest of its compositional elements. A heavily shelled (but still standing) multi-storey building, with broken balcony fences, shutters hanging down from windows, and flames coming out from the upper floors, is depicted with anthropomorphic features: it has limbs and a face, and is bleeding; is wearing army boots, and appears to be running. In one hand, this anthropomorphic building holds a torch, on top of which a flame is burning, its flames forming the Arabic word *Lubnān* (Lebanon). On the side of the torch, is another Arabic word: *al-imān* (the faith). In the other hand, leaning against one of the building walls, is a map of the nation state of Lebanon. The cedar tree depicted on the Lebanese flag is present on the map, where we can also read the phrase *al-qadiya al-lubnāniyya* (the Lebanese question). The building is also stepping onto an indistinct and 'antlike' (Maasri 2008, 103) hostile crowd, from which missiles or spears are being thrown. The Arabic text under the crowd reads a*l-qiyāl al-hamajiyya al-'arabiyya* (the barbaric Arab tribes).

There are several geopolitical notions overlapping in the poster: their meanings and implications are renegotiated and reinscribed onto the urban built environment of the neighbourhood, thus contributing to establish a new urban geopolitics, where divisions and hostility at the neighbourhood scale intertwine with the wider geopolitical realms of the nation and the Arab region.

Particularly telling is the depiction of Ayn al Rummana's role in the conflict: a key space and presence in the city, embodying steadfastness, literally carrying the torch of national faith, and the burden of the Lebanese sovereignty question on its back. This happens while 'the world is sleeping', a painful reminder of the geopolitical script of non-intervention which allowed the city to become partitioned and the proliferation of these urban geopolitical discourses of hostility and 'us vs them'. The materiality of the mortar and concrete of the building and even the objects of furnishing – balustrades, shutters – all are enrolled into a precise geopolitical script: maintaining Lebanon's territorial sovereignty against the territorial project of pan-Arabism and preserving Lebanon as a sovereign nation within the borders of the independent state in

1946 and, before that, the 1920 borders of the State of Greater Lebanon under French Mandate. This geopolitical rationale, advocated by a nonstate armed organisation, draws directly from a mainstream geopolitical narrative with a clear colonial genealogy. This is a subjugated geopolitical knowledge, that is differential, yet speaking directly to the geopolitical mainstream. It is nonstate, yet drawing from the established power fundamentals of the state. These rationalities function as a stark counterpoint to the depictions of Lebanon's militias as 'undisciplined elements' 'eating away' Lebanon's sovereignty found in the official diplomatic accounts we saw in Chapter 5.

Affirming group identity via hostile representations of the other was a common practice in the visual political propaganda from the civil war, where 'derogatory representations of the enemy in traditional attire have been employed to reinforce the opponent's otherness at the level of ethnicity and to present it as belonging to a backward and undeveloped community' (Maasri 2008, 103), and also across other forms of militia statements, as we saw earlier in Koleilat's appeal to the state to stop the 'fanatics' from attacking the city centre. In the poster, the cause of Lebanese nationalism and its hostility both to the Syrian presence via the ADF and, more specifically geopolitically, to the territorial project of pan-Arabism understood as threatening Lebanese sovereignty and boundaries, is expressed using clearly orientalist tones to depict the Arab crowd as chaotic, backward, and hostile.

New territorial knowledge and ways of understanding and controlling the city developed among the militias. A congested built environment mainly lacking master planning, and developed by private investors in the laissez-faire financial climate of the prewar era (Kassir 2010), had now become a controllable terrain for numerous nonstate actors via that same 'inverse geometry' (Weizman 2007, 189) considered by scholars as typical of the rethinking of urban territory by the theories and practices of 'new' asymmetric warfare (Graham 2005; Kaldor 1999). This includes the technique of 'walking through walls', widely used by the IDF's contemporary urban manoeuvres in the Palestinian occupied territories (Weizman 2007), which a Edouard recounted while describing the battle for a major frontline in Beirut's city centre:

> They open[ed] holes through the buildings so [the distance] it's very short, you know, you start from near Qantari and then you get to l'Orient le Jour [cutting through] the whole building in few hundred meters, because when you cut the interior of the building, between walls, you have an extraordinary force, and you are indoors. (Edouard)

Describing the Beiruti residents' coping strategies during the early street-fighting phases of the civil war, French geographer Michael Davie

(1983) remarked how the street fight blurred classic distinctions between public and private space—for example, by exposing balconies to the dangerous outside and by reclaiming secluded public back alleys as part of domestic secure space. Perforating domestic and private spaces for military purposes goes even further. It dissolves the meaning of inside and outside, continuously recreating the city as a flexible and three-dimensional tool of violence.

While the territorial trap of the geopolitical scripts in the official diplomatic circles took the state as exclusive interlocutor and interpretive tool of the conflict and dismissed the militias as essentially fanatics, on the urban ground a series of differential geopolitical knowledges were taking shape and being enacted on the urban ground of a partitioning city. These knowledges relied on specific architectures of enmity, othering the enemy, and recasted and renegotiated mainstream geopolitics into the scale of the neighbourhood, enrolling the built environment in a battle that was much about urban territory as it was about the wider geopolitical role of Lebanon. It is to the specific role of urban space and the built environment as a geopolitical agent that I will now turn.

## Geopolitical architectures: The battle of the hotels

The locations of the most furious battles for the possession of the city centre were often initially the result of tactical calculations in order to gain more terrain or to prevent adversary militias from advancing, as Edouard explained:

> we cannot say that there was an initial decision to destroy the economy of the country, or to destroy those sites that were considered as being a symbol, or economic poles, or whatever. Initially, it was simply about fighting and that could not always be controlled. (Edouard)

The shelled skeleton of the Holiday Inn hotel stands on the western edge of Beirut's city centre, a few metres back from the seafront, in the neighbourhood of Kantari. The hotel was part of the wider St Charles complex, named after its location on the site of the former Hospital of St Charles Borromeo, which was founded in 1908 and subsequently relocated to Baabda, in the city's hilly outskirts. The hotel became one of the emblems of Lebanese modernist architecture, designed by French architect Andre Wogensky, a student and successively a collaborator of Swiss-French architect Le Corbusier. Wogensky collaborated in the design of the building with Lebanese architect Maurice

Hindieh. Although adapted architecturally to Beirut's location and climate, Beirut's Holiday Inn embodied the western (and especially American) values of functionality and modern technology, although in a less grand and luxurious style than the Hilton chain (Wharton 2004).

But between October 1975 and March 1976, the city centre became the focus of the battle for the partition of Beirut. On 27 October 1975, the unfinished Murr Tower, in the vicinity of the hotel, was occupied by the Al Mourabitun[3] militia, while the area with the Holiday Inn and Phoenicia hotels was still held by the militias of the Lebanese Front. Intense fighting broke out between the Murabitun and a division from the Kata'ib militia that had barricaded itself in the Holiday Inn. On 15 December 1975, the eighteenth cease-fire agreement of the war ordered the militia evacuation from all high-rise towers and hotels, before the army took charge of downtown. But in March 1976, fighting returned, this time concentrated on a front with 'the hotel Holiday Inn and Hilton and the Starco shopping centre on one side [and] the hotels Phoenicia, Saint Georges, and the Murr Tower on the other' (L'Orient le Jour 19 March 1976, 1). In what was a veritable battle for the occupation of buildings' heights in order to control the territory below, the Holiday Inn, again occupied by the Phalanges militia, became the object of a fierce confrontation before eventually falling to the National Movement on Monday 21 March 1976.

Beirut's Holiday Inn is not the first hotel to be targeted and taken over by combatants – a famous example is the Habana Hilton in Cuba taken over by the Communist revolutionary movement (Lisle 2016; Merrill 2009). Strategically, the battle of the hotels (*Ma'raka al-fanadiq*), and especially the fighting over the Holiday Inn, was a crucial moment in defining the partition of the city into two sectors and in stabilising the last stretch of the main confrontation line (*Khatt at-tamas*, popularly known as Green Line):

> The battle of the hotels was fundamental because it would complete the [Green] Line [...] Beirut became a unique line of fire [...] it started in Kantari, until it reached the Holiday Inn and the Starco building where the confrontation line was completed [...] and it became a stable line. (Omar)

The green line had started from the suburbs of al-Shiyyah and Ayn al Rummana in April 1975, gradually progressing north 'along the Damascus highway through Martyrs Square to the west gate of the harbour' (Hanf 2015, 215) and was eventually completed, having reached the sea, as a result of the battle of the Holiday Inn. The battle of the hotels, therefore, and especially the showdown between National Movement and Lebanese Front around and inside the Holiday Inn hotel in the Saint Charles building, is a highly important node within the urban geopolitics of the Two Years' War. In the regional panorama of the time, it can be interpreted as a strikingly three-dimensional 'edge [...] of bipolarity where American- and Soviet-led spheres of influence conflicted each other,' (Lisle 2016, 125). In Holiday Inn's case, the conflict was between the Socialist/pan-Arabist alignment of the National Movement (with logistic support from a number of Palestinian armed groups) and the Lebanese Front aligned with the socio-economic values of the US bloc. The physical presence of the Holiday Inn grounded the showdown between these conflicting views of the world.

> Before the war, the best cinema was the Saint Charles, and we used to attend it; but we did it seeing it in an anti-class logic [...], as this cinema was in fact a celebration of the capitalist system. During the war, this was also joined by the value of the Holiday Inn as a fortification, whose seizing was important to hit the moral of the enemy. The moral defeat was on two sides: on the moral of the enemy and on the capitalist, bourgeois system, which we associated with the opposing part. (Omar)

The physical concentration of high-rise buildings known as the hotel district, was a fruit of the investment of private capital, both Lebanese and international (mainly Western) that, between the 1950s and mid-1970s, profoundly transformed the surrounding costal neighbourhoods of Aïn el-Mreisseh and Minet el-Hosn and their social practices (Sawalha 2010). The presence of embassies, luxury hotels, furnished apartments for businessmen traveling between Beirut and the Persian Gulf, as well as the many night clubs, and restaurants in the alleyways around the hotels, constituted a cosmopolitan space that, however, relied on unequal interaction with Beirut's increasingly alienated working classes, who gradually came to see the hotels as chimeras unavailable to them. In the words of a local architect, 'The Grand Hotels ... were not linked with the rest of the country, but with people of the big capital and with those people who became occidentalized to the bitter end'. For Abu Layla, the identification of the hotel neighbourhood with the international capital is present, but secondary to the tactical value of the hotel as a territorial asset for the militia and the sectarian makeup of Beirut:

> [the hotel area] was a new Lebanon that was for the rich, not for us. Despite this, every time that a tower fell in the hands of our enemy, we – the Christians of the Eastern part of Beirut – felt that they were approaching, that they wanted to kill us, that they are now closer to us. (Abu Layla)

The battle of the hotels involved both tactical calculations by the militia, and specific geopolitical narratives. Paramilitary tactics and practices, and less tangible aspects such as class and sect, were at once negotiated through the physical spaces and built environments of the battle. The militias 'disputed the city between themselves' (*L'Orient le Jour* 1975b, 1) not purely on the tactical/territorial level, but by inscribing geopolitical narratives onto space and the built environment, and contesting them through it. When the battle ended, with the fall of the Holiday Inn to the National Movement fighters on 23 March 1976, a spokesperson for the Murabitun militia defined the battle to 'conquer' the Holiday Inn as one 'that has an aim which is at once military and political' (*L'Orient le Jour* 1976c, 4).

As these spatialities were being reworked by political violence, the built forms became at once the tool *and* the product of these particular practices of war, and the lived spaces of urbicide were enacted not onto but *through* the city's buildings, transforming their peacetime functions and appearance into tools for warfare. A journalist who covered the hotel battle in 1976 recounted that

> when the Palestinians and the Murabitun decided to seize the Holiday Inn [...] they were shooting bullets systematically against the hotel, floor by floor. While bombing one of the floors, it went up in fire and as the hotel was equipped with some sort of automatic fire alarm, the fire activated it and the water helped to put off the fire [...] and so it was the water and this electronic system that was specifically made for the customers first of all, through which the hotel tries to defend itself against the warrior. (journalist, 25 October 2005)

Oppositely to the geopolitical narratives developing around Ayn al Rummana, the battle of the hotels and especially the fall of the Holiday inn, embodied and indeed helped to amplify the geopolitical imaginations around the politico-territorial project of pan-Arabism (Mufti 1996; Eyal Zisser 2003), of a Lebanon with a strong Arab identity and spatial connection to its Arab hinterland that the National Movement supported. This is reflected in the words of Kamal Younis, the leader of the Socialist Arab Union militia, who first organized the assault to the hotel, who stated at a press conference that

> the Holiday Inn has fallen 'for the safeguard of the Arab belonging of Lebanon' [...] The martyrs have fallen for the 'national cause' [...] We affirm once again that Lebanon will remain Arab and that all its children, its institutions, its army and its culture will play their role by serving

the Arab causes [...] We are building the Lebanon of the future as all the martyrs that have fallen in the last months wanted it, a democratic Lebanon, where all the citizens will be treated equally, and where social justice will be created especially for the forces of the working classes. [...] The army of Arab Lebanon [is the] image of the Lebanese army of the future, of national and non sectarian Lebanon. (L'Orient le Jour 1976b)

This position was echoed by Al- Mourabitun whose spokesperson stated to the press that 'the aim of our people is the protection of the Arab identity of Lebanon and the realization of a free, socialist, and united society' (AL-Nahar 1976, 3).

The battle of the hotels was of crucial tactical value for dividing Beirut through completion of the Green Line. But partitioning Beirut was also the embodiment of the extreme confrontation between two opposed territorial projects: that of Lebanese nationalism (on the Lebanese front's side) and pan-Arabism (on the National Movement's side). The battle of the hotels, and especially the emblematic fall of the Holiday Inn, was thus made the subject of celebration, commemoration through press conferences, communiqués, and visual propaganda, again through the use of posters.

I want to concentrate on two posters released by Al Murabitun militia. The first (Figure 7) commemorates the martyrs fallen in the attempt to take over the Holiday Inn; the second (Figure 8) was released in 1977 for the first anniversary of the hotel battle. Both posters report al-Murabitun leader Ibrahim Koleilat's statement at the end of the battle: 'on 21 March 1976, Al-Murabitun crushed the symbol of the fascist treachery, and swore that they will continue the fight whatever the price'. The Holiday Inn has become here, we read in the Arabic text, 'the symbol of the fascist treachery'. Fascist – in this perspective – are the isolationist militias.[4] They were considered treacherous by the National Movement because of their 'isolation' and refusal to take into consideration the Arab identity of Lebanon and instead insist on an idea of nationalism deeply connected to the borders established for Lebanon in 1920.

Visually, the perspective from which the hotel is depicted is the same in both posters: it represents the side of the hotel facing uphill towards the hill of Kantari, where the National Movement militias were positioned. It is from this side of the building that the tactical manoeuvre allowing the National Movement militias to break into the building was conducted:

As the Murabitun could not execute this plan by themselves and capture the hotel, they decided to do the job under the leadership of Ahmad Jibril's PFLP-GC. Jibril was an expert in explosives. He and his men blew up a

**Figure 7** Al-Murabitun propaganda poster. American University of Beirut, Archives & Special Collections. Political Posters. Poster No. 158-PCD2081–17.

wall that separates the Holiday Inn from the Phoenicia, and from this open wall the combatants – the majority was Palestinian – entered in the Holiday Inn and the Lebanese Forces – the Phalanges – began to leave the place. (journalist, 25 October 2005)

This manoeuvre procured Ahmad Jibril and the National Movement the attention of the media: press conferences followed, where the National Movement's spokesmen released their declarations. Through this event, the power – and the geopolitical imaginations – of the militias of the National Front could be communicated, through the media nationwide and turned the battle into a spectacular event. The first poster includes images of fighters posing at the end of the battle and during the press conference held by Ibrahim Koleilat. As a journalist who covered the battle in 1976 told me, Ahmad Jibril carefully planned the exposure to the public eyes of a battle for one building that was not only tactical and instrumental for the division of the city, but crucial for the affirmation of their specific geopolitical imaginations:

> It was [...] about midnight and his [militia representatives] phoned me and asked me if I can send a photographer [...] I sent him a photographer and [Ahmad Jibril] insisted to have the copies of the pictures before they got published. [...] He smartly considered that passing these pictures through the newspapers meant to give everybody the proof to the whole world that it was Ahmad Jibril that had destroyed the hotels. (journalist, 25 October 2005)

The second poster (Figure 8) reports the same statement by Ibrahim Koleilat, whose portrait is in the upper left corner. Here, the overlapping of considerations about tactical and territorial superiority and of assertion of specific geopolitical narratives is more obvious. The logic of urbicide, as a process of re-definition of geopolitical imaginations via the reconfiguration of urban space, is illustrated in the graphic of the Al Murabitun fighter smashing the Holiday Inn with the butt of his rifle. Urban destruction in Beirut did not happen in a void; neither was the very act of destruction an end in itself. Rather, it was a process of redefinition of existent spatial order of the city. In the National Movement's view, Al Murabitun fighters defeated the Lebanese Front militias and won the battle of the hotels by evicting them from the Holiday Inn. Through this act, not only did Al Murabitun and the National Movement – strike a tactical victory, but they also affirmed their geopolitical imagination onto the space of Beirut, by seizing what had become a major paramilitary asset (the hotel), sealing the Green Line, and finally claiming control on specific urban territories.

The sentence by Koleilat quoted in these two posters was also reproduced in graffiti on the walls of West Beirut (Chakhtoura 2005) and remained part of the urban landscape well after March 1976 and into the year-long

**Figure 8** Al-Murabitun propaganda poster. American University of Beirut, Archives & Special Collections. Political Posters. Poster No. 159-PCD2081–16.

truce after the entry of the ADF into Lebanon in 1977. The hotel and the statements related to it gained a repeatable materiality: while linked to the material 'conquest' of one building, they also embodied a specific geopolitical knowledge that lived beyond the takeover of a building and was mobilised more widely within discourses about Lebanon as a nation state and its role in the political map of the region. Holiday Inn became an urban and geopolitical icon: its shape recognisable, its presence strategically crucial for the completion of the green line, and embodying the specific territorial and political project of the National Movement and its pan-Arabist agenda.

The Holiday Inn, populating pre-war everyday life with its modern and spectacular architecture, had now become both a strategic position and a political symbol.

On the fighting lines that ignited across the suburbs and the historic centre of Beirut in the first two years of the civil war, the irregular militias disputed, renamed, repopulated and effaced various portions of the city. They fought with increasingly heavy weaponry that they were trained to master, often by army members, from before the start of the war. The militia organization became increasingly structured, their geographical and technical knowledge improved and the material imprint of their activities became extensively visible. But the militias did not operate in a space exclusively composed of material objects; they also produced geopolitical narratives about the nation state, power and territory. These narratives were mobilised through representations of and action upon specific portions of the urban fabric (like Ayn al Rummanah and the Holiday Inn). The built environment became a constitutive part of the production of geopolitical meaning, and also bore the material consequences of that production, as specific built environments became targets for reasons beyond military necessity. Looting, for example, was rife especially in the hotel area, where, after the battle was over '[the combatants] began to empty the Holiday Inn of everything but the concrete: the tiling, the sanitary furniture, everything that was inside and wasn't wall. [...] carpets, mini-bars, pieces of furniture and so on. They left nothing' (journalist, 25 October 2005).

Those same militias, however, were depicted in the official statements of governments in Europe, the United States and Israel, as irrational and irresponsible forces, placed outside politics, who had caused Lebanon to deviate from its tradition of coexistence and plunged the country into chaos. And yet, the urban militias produced their own discourses about power and its specific space in the city, despite remaining invisible in official diplomatic discourses about intervention and reconciliation.

As the war continued, a constellation of new urban practices emerged, including provision of security and basic services such as water, energy and sanitation (Harik 1994). At the end of the Two Years' War, in 1977, 142 militia Popular Committees operated in Beirut, employing 1,400 civil servants for a population of 250,000. These committees 'generally paralleled that of the state's service agencies' (Harik 1994, 16), and were active in coordinating repairs to damaged buildings and infrastructure, refuse collection, and sheltering displaced persons. These localized arrangements and power practices did not enter the narratives of official international foreign politics, but they drastically transformed the spatiality of thousands of city dwellers. Spatiality is defined as the relationship between people and things (Kobayashi 2017), meant as the capacity of space to shape specific social interactions. Urbicide, then, acts against the urban built environment in order to change the relation between people and things in the city. This change is not an end in itself, but is part of a renegotiation *at ground level* of wider geo-territorial projects that are often inspired by established geopolitical codes – as with the case of Lebanese nationalism and pan-Arabism in Beirut.

This chapter constituted the second part of a contrapuntal reading of the geopolitics of the Two Years' war in Beirut. It has departed from the state-centred, disembodied and primordialist scripts of official geopolitics seen in the last chapter and illustrated instead how militias fighting in the city engaged with mainstream geopolitics to envision differential and highly contested geopolitical imaginations for Lebanon. Contrary to the claims of official diplomacy about Lebanon being lost in the hands of fanatic and irresponsible actors, this chapter has shown how armed non-state actors mobilized and concretized differential geopolitical narratives and knowledges through the reconfiguration of urban space. These are not truer or better representations: 'this is no simple inversion of power in which an "authentic" voice from below speaks effectively back to power' (Lisle 2016, 292), but a way to see how these encounter each other and (re)shape urban space. The architectures of enmity taking shape in the early 1970s had grounded themselves *in and through* urban space, reinforced by the stabilization of the main confrontation lines within the city.

Although sovereignty and territoriality constitute the pillars of international law, critical political science and IR scholarship has analysed how nation states negotiate and arrange different de facto combinations of sovereignty (Biersteker and Weber 1996; Clapham 1998; Krasner 2001, 1999). Political geographers have also long highlighted the inadequacy of the Westphalian model of state sovereignty to cope with globalisation (Agnew 2009) and contexts where nonstate actors and polities ought to be accounted for (Fregonese 2012a; McConnel 2009) in order to gain a contextually richer understanding of how sovereignty actually works.

Far from disappearing, the State operated well into the war. In 1975 and 1976 parliament reunited in a peripheral location, diplomatic contacts continued (Hourani 2010b), the president addressed the nation and the army existed, although acting mostly in self-defence and often being targeted by militias of all sides. Rather than plunging the city into chaos, nonstate armed actors articulated forms of sovereignty that coexisted and even blurred with the state. The militias acquired increasingly state-like functions, while the army and elements of the government associated and embedded themselves within militia networks. These 'tight circular connections' (Graham 2004d, 23) between the state and nonstate have far from disappeared in Lebanon and in Beirut. On the contrary, as the next chapter will show, they continue to shape the way in which security, sovereignty and conflict are understood, conducted and resolved.

# 7

# Beirut's Hybrid Sovereignties: The May 2008 Clashes

This chapter tackles more centrally the question of sovereignty, urban space and war. It fast-forwards to another moment of intra-state conflict in 2008, when political polarization in the aftermath of former Prime Minister Rafiq al Hariri's killing in February 2015 and the consequent wave of protests known as the 'Cedar Revolution' led to clashes between armed groups in and around Beirut and surroundings.

In May 2008, exactly fifty years from the assassination of Nasib Metni that sparked the street battles that led to Operation Blue Bat in 1958, several neighbourhoods in Beirut, as well as in Sidon, towns and villages in the surrounding Chouf mountains (Aley, Aytat, Kayfoun, Baysour, Shuweifat), and the Bekaa Valley (Bar Elias), experienced rounds of deadly clashes – what is popularly known as 'May 7th'. Several neighbourhoods in Beirut were engulfed by violence between 6 and 12 May. Mainstream media and diplomacy portrayed the clashes as the proxy stage of a bigger 'real' geopolitical battle between Saudi Arabia and Iran. An urban geopolitical approach highlights instead how the 2008 conflict in Beirut involved concrete urban infrastructures of communication, transport, information and security, in ways that were not always the direct derivation of those bigger regional geopolitical struggles, but rather – and very much similarly to the Two Years' War – a localized renegotiation of those struggles through the physical space of the city. The chapter also interprets these urban geographies observing how power over urban territory and infrastructure was negotiated in concerted ways between state and nonstate actors. To do this, I propose the concept of *hybrid sovereignty*[1] as a more effective way than traditional Westphalian approaches to sovereignty to make sense of Lebanon's de-facto power arrangements that cut across normative divides between state and nonstate, especially in moments of conflict.

Despite the abundance of critical research on sovereignty in disciplines such as geography, political sciences and anthropology (Agnew 2009; Biersteker and Weber 1996; Krasner 2001), much of foreign policy theorists and practitioners still rely on a state-centred relist view of global politics. The state is still the main tool to narrate territory and mainstream approaches place nonstate actors in a realm of illegitimacy that are morally opposed to the state sphere. Without, for a moment, justifying the violent actions of unrecognized armed actors, the hybrid sovereignty perspective presented in this last chapter offers a pragmatic view to interpret and understand how, in conflict, power does not dissipated, but is instead redeployed and redistributed beyond the normative frame of legal sovereignty. Hybrid sovereignties also allow us to account for and act on those spaces where hybrid articulations of sovereignty actually shape everyday life and material networks in conflict cities.

## From lack of sovereignty to hybrid sovereignties[2]

Political scientists and political geographers have treated sovereignty as a social construct that is 'neither inherently territorial nor ... invariably state-based' (Agnew 2009, 9) and as a process resulting from 'knowledgeable practices by human agents, including citizens, non-citizens, theorists, and diplomats' (Biersteker and Weber 1996, 18). Sovereignty enactments that are other than territorial include, for example, the idea of shared sovereignty in the Israeli/Palestinian peace process (Mavroudi 2010) and the everyday, nondramatic 'tacit sovereignty' practices of the Tibetan Government in Exile (McConnell 2009).

It is worth reviewing three approaches that provide different modes of discussion of sovereignty. The first is Robert Jackson's (Jackson 1986) notion of 'negative sovereignty', whereby a state is internationally recognized yet defective in domestic sovereignty. Jackson emphasizes control of the capital city as crucial to retaining international recognition. This interpretation can overlook where and how control of territory is actually enacted due to its view of 'the conventions of diplomacy as a kind of international board game, in which control of the capital counted as having "won" ' (Clapham 1998, 50–51). Even if we consider Lebanon as a state of negative sovereignty, by focusing on the institutional centre of the capital Beirut we overlook other, less dramatic, practices of sovereignty in the greater urban sphere and beyond the capital.

The second approach is James Sidaway's (2003) idea of 'sovereigntyscapes', in which states (especially postcolonial ones) considered weak are not necessarily lacking or leaking sovereignty but experiencing the participation of many actors in continually negotiated sovereignty formations. In other words, there is an excess rather than a lack of sovereignty. Sovereignty therefore is determined not simply by either absence or presence 'of connection, power and capital, but by a particular form and experience ... of these' (160). Sidaway's view of sovereignty is not state led or necessarily territorial, and opens 'the possibility of other analytical frameworks, beyond the issue of more or less sovereignty, beyond the presence or absence of undifferentiated sovereign power, towards a contextual understanding of different regimes, apparatus expressions and representations of sovereignty' (174).

The third approach is John Agnew's (2005; 2009) idea of 'sovereignty regimes'. With this concept he attempts to make sense of sovereignty in times of globalization. Agnew accounts for the complex ways in which states enact different 'combinations of degrees of central state authority and consolidated or open territoriality' (Agnew 2005, 456), instead of counterposing the two sovereignty regimes. Agnew aims to overcome the dialectics between de jure and de facto sovereignty by developing the idea of 'actually existing or effective sovereignty' at work besides state authority (Agnew 2005, 456), and by adopting a Foucauldian notion of power as decentred and locally reproduced rather than emanating from the top down from a unique institutional source (2009). Agnew goes beyond the work of Jackson and Sidaway by declaring that 'de facto sovereignty is all there is' (Agnew 2009, 7).

Negative sovereignty, sovereignty excess, and sovereignty regimes all stress the multiplicity of actors negotiating state territoriality; less apparent in these ideas, however, is the challenge to the discreteness of these actors. I want, instead, to build on these approaches and focus on the cross-contamination of different state and nonstate actors, to the point that the state and the nonstate become difficult to distinguish and, instead, new entities are produced. Hybrid sovereignties are the result of this cross-contamination.

Hybridity is a mixing of different ontological categories, 'a condition describing those things and processes that transgress or disconcert binary terms that draw distinctions between like and unlike categories of objects such as self/other, culture/nature, animal/machine or mind/body' (Gregory et al. 2009). Among the intellectual projects in which hybridity finds space are cultural and postcolonial studies on identity and transculturation, such as that by Homi Bhabha, for example, who describes hybridity as a perspective that goes beyond mixing into the production of new entities: '[Hybridity] overcomes the given grounds of opposition and opens up a space of translation: a place of hybridity [...] where the construction of a political object that is new, *neither one nor the other*, properly alienates our political expectations and changes, as it must, the very form of our recognition of the moment of politics' (Bhabha 2004, 37).

Science and Technology Studies (Haraway 1991; Latour 1993) uses the notion of hybridity to counteract dialectic ontologies about mixing as the mere sum of difference, and to dissolve binaries between categories like nature and culture, including the 'nonhuman' in ethical and political debates. Hybridity as a dissolution of binaries and categories also finds space in biophilosophy with its consideration of bodily practices and feelings as epistemologically equal if not more valuable than cognitive processes, and in debates on the practices of the everyday in which the mundane is considered a legitimate form of pragmatic and performative knowledge beside and in relation to official and formal knowledges. Hybridity joins worlds that are discursively separate, as the 'exercise [of] other modes of travelling through the heterogeneous entanglements of social life' (Whatmore 2002, 3) and uproots binaries, 'disturb[ing] the territorialisations of self, kinship, neighbourhood and nation and invit[ing] other "languages of attachment"' (167). Hybridity is, most importantly, also a political project as it proposes a 'performative' view of territoriality that considers 'the tangle of socio-material agents and frictional alignments in which it is suspended' (87) and reaches different conceptions of and possibilities for governing territory. This links the idea of hybridity to the discussion of sovereignty and makes the case for bringing hybrid epistemologies into the study of political power and territory. An embodied, hybrid epistemology involves a close analysis of the 'labours of division' implied in the performance of territoriality and 'the host of socio-material practices—such as property, sovereignty and identity—in which they inhere' (6). Epistemologies of hybridity allow the repositioning of accepted international and local discourses about Lebanon 'lacking' sovereignty. Rather than a 'weak state' lacking sovereignty, we should see Lebanon as a constellation of hybrid sovereignties. These epistemologies highlight the intimate connections between state and nonstate, and the role of urban space as coconstitutive of sovereignty.

## Beyond the state/nonstate dialectic

Hybridity allows us to unpack traditional international relations binaries that maintain state and nonstate actors in different spheres of legitimacy regarding the administration of political violence. The 'control of the means of internal and external violence' is part of the 'late-European' practice of tracing fixed boundaries around the 'realm of administration of the state' (Giddens 1981, 190). The state is still the 'omniscient narrator of territory' (Ó Tuathail 1996b, 12) for mainstream approaches that condemn nonstate actors as 'directly opposed' (Elden 2009, xxii) to the state sphere. This binary places state actors (for example, regular armies) in a realm of legality and 'legitimate war' and places nonstate actors (for example, irregular combatants) in that of illegality and 'terror' (Elden 2009; Gregory 2006). These 'structures of legitimacy' (Gregory 2006, 100) underlie the 'rhetorical spaces of late modern war' (2010, 170).

These are critiques of those 'zero-sum' views that conceive either a completely and homogeneously deployed state sovereignty throughout a territory or the total absence of political organization. Increasingly, however, a 'crisis of representation' (Sidaway 2003, 159) in international sovereignty is exposing the disjuncture between the conventions of global politics and the actual practices of sovereignty, as it gets enacted in different ways in different contexts (Agnew 2009). Increasingly, these negotiated practices are not only visible but also internationally relevant: polities not coinciding with the state but implementing state practices could be 'a valuable glimpse of possible geopolitical futures' (McConnell 2009, 12). Breaches in the state's monopoly of political violence emerge in the 'marriages of military convenience' (Gregory 2010, 170) between regular and irregular armies: coalition forces brokering deals with insurgencies in Baghdad, the US Army integrating Afghan militias into military support groups, alliances between security forces and drug cartels at the Mexican borderlands, and Hosni Mubarak resorting to paid thugs (baltajiya) to thwart the revolution in Tahrir Square. It is also increasingly apparent that the challenges of late-modern conflicts to classic configurations of sovereignty—especially the renegotiation of the role of state nonstate actors—are often articulated around the built environment of cities.

## Sovereignty and the built environment in the May 2008 clashes

Between 5 and 12 May 2008 several neighbourhoods in West Beirut experienced armed clashes between opposing militants of the '14 March' and '8 March' coalitions (Figure 1). On 6 May the government outlawed Hezbollah's private communications network for violating Lebanon's sovereignty, and dismissed the security chief at Beirut international airport, General Wafiq Shuqeir, who was allegedly close to Hezbollah. On 8 May, in a televised speech, Hezbollah leader Sayyid Hassan Nasrallah compared the government move to a 'declaration war against the resistance'. On the same day, in parallel with a national strike by the Lebanese General Workers Union calling for a minimum wage increase, opposition groups blocked roads in and around Beirut, including the highway leading to Beirut's international airport. Fighting continued for the following week and several areas in West Beirut were controlled by armed militants linked to the 8 March coalition. While scholarly analysis of the 2008 clashes is still limited (however, see Harris 2008; Shehadi 2008; Stel 2009), the international media consistently represented those events as the closest the country had come to the 1975–1990 civil war (Le Monde 2008; McElroy 2008; McLeod 2008; Muñoz 2008). There was an abundance of headlines like, ' "Shock and awe" in Beirut' (Harb 2008) and 'Lebanon on the brink of chaos' (Corriere della Sera 2008).

Numerous political analysts and security experts at the time framed these events through realist regional geopolitics, seeing the clashes as a struggle for control between Iran, Saudi Arabia, and the West (Naharnet 2008). Conversely, other commentators, as well as many of my research participants, explained the May events in terms of Beirut's everyday urban life and infrastructure, especially those elements related to security and communications.

At the start of each of my twenty interviews conducted between October and December 2010, I posed the same question: 'what made you realise that there was a problem?' Almost all respondents referred to three main events: the government investigation into Hezbollah's mobile communication network and especially its security installations at Beirut international airport; the general strike of 6 May, during which roadblocks were created and tyres burnt; and Nasrallah's speech on 8 May, which defined the government's attempts to dismantle Hezbollah's private mobile communication network as the 'crossing of a red line' (Al Jazeera English 2008).

Middle East analyst Nadim Shehadi (2008, 13) emphasizes the importance of the events leading up to the clashes, particularly the airport incident: 'the government challenged Hezbollah's control of security at Beirut airport and launched an investigation into security cameras on the runways and on the group's bases. Hezbollah's leader, Sayyid Hassan Nasrallah, said this was exposing its defence capabilities and considered it tantamount to a declaration of war.'

After remaining almost uncontested since the end of the civil war, the material infrastructure of Beirut's southern periphery, where the airport is located, became the centre of controversy. Attention was focused on some containers stacked in an empty square a few metres from runway 17, accessible from a non-asphalted road within the Hezbollah-controlled neighbourhood of Ouzai. On 3 May Joumblatt, at the time one of the representatives of 14 March, denounced to the press the presence of these '100 and 1 containers', one of which he alleged was equipped with Hezbollah-owned surveillance cameras. Government representatives accused Hezbollah of preparing a major attack against politicians' jets landing on the runway, and the issue was covered extensively in the media. New TV aired a documentary using satellite images of the runways to show the different fields of vision of the airport's legitimate security cameras and of Hezbollah's 'private eyes'. Different regimes of vision were cast on the infrastructure of the airport, which became a terrain for practices of hybrid sovereignty: owned by the state but simultaneously tolerating nonstate 'observant practices' (MacDonald, Hughes, and Dodds 2010) alongside those of the state.

The value of urban infrastructure in the control of territory in Beirut cannot be appreciated without considering Hezbollah's sovereign practices on certain portions of the city. Hezbollah retains effective sovereignty in the southern suburbs of Beirut and the south of the country (Norton 2014). The former militia has presented itself in the post-civil-war arena not only as a political party and armed resistance (Kramer 1993; Majed 1996; Norton 2014; Zisser 1997) but also as a social provider of infrastructure. Hezbollah provides housing, hospitals, water, and gas in the Shi'a-majority area of south Beirut, al-dahia aj-janubiyya.

This area is an informal settlement whose administrative and legal boundaries are indefinite (Fawaz and Peillen 2003; Fawaz 2009), and whose ownership and citizenship makeup are not entirely managed by or informed by the state sphere (Bou Akar 2005; Roy 2009). The provision of public services allows Hezbollah to reconcile its multiple roles through the notion of mujtama' al muqawama or 'society of resistance' (Qassem 2012). The masterplan for the reconstruction of the southern suburbs started after the 2006 war with Israel and the appropriation of historic memory and its infrastructure of memorials, museums, and themed parks in south Beirut and South Lebanon in collaboration with the Ministry of Tourism (Harb and Deeb 2009) constitute two examples of this provision for the population. Hezbollah is, therefore, simultaneously a political party (part of the government today but part of the opposition in 2008), an armed resistance movement, a provider of social services and a provider of infrastructure: it is simultaneously part of the state, nonstate, and state-like.

Another way in which the built environment became entangled in the expressions and contentions of sovereignty is the deliberate targeting, from the first days of the clashes, of March-14-aligned TV stations like Future TV and newspapers like Al-Mustaqbal. This 'switching off' of the pro-government media infrastructure was a carefully planned response by the Hezbollah-led opposition to government moves to close its phone communication network. In particular, the Al Mustaqbal newspaper building came under attack with heavy weapons such as rocket-propelled grenades. Writer and newspaper editor Rami Khoury (Khoury and Garfield 2008) said Mustaqbal was 'very narrowly and clearly targeted'. Empty buildings under construction were used as platforms for the launch of rocket-propelled grenades and for automatic weapons, which suggests a degree of preparation during the previous nights (interview with journalists, Beirut, 24 November 2010).

This was an attempt to destroy the infrastructure itself: newspaper staff told me that weapons fire was aimed at the interior rooms of the building at night when there was only a handful of journalists inside (interview with journalists, Beirut, 24 November 2010). Eventually, two floors of the building caught fire, including the archives. Preparation, professionalism, but most importantly coordination between the army, Hezbollah, and private security firms on media premises prepared the ground for the implementation of hybrid forms of sovereign practice. Paul Cochrane (2008) describes this coordination between state and nonstate elements in Hezbollah's attack on Future TV. He describes the departure of the Lebanese Army, the local 14 March militia, and the private security guards, then the arrival of Hezbollah men and the coordination between the army and Hezbollah in evacuating the building and switching it off in a professional manner. On the morning of May 9, the FTV employee said an army officer entered the offices.

> He said [Hezbollah] armed men were outside and 'if you don't leave the building, they will come in or burn the building down'. The news editor asked for … employees to be allowed to leave and the station to not be harmed, as well as for one technician to stay behind.
>
> After everyone left, according to the technician, the Colonel came back with Hezbollah technicians to be taken to the master control room. Cables, uplinks and satellite links were cut. They were professional and knew what they were doing. They needed to find the server, so made the technician call the head technician to find out, and on the phone [Hezbollah] said they knew where he lived. (2008)

In the next few days the army froze the measures taken by the government against Hezbollah's communications network and urged all fighters to withdraw from the streets. On the night of 10 May, the 8 March forces, defying local and international alarm over an imminent coup d'état and with little resistance from the pro-government activists, handed over the occupied neighbourhoods to the neutral control of the army (Martinez and Volpicella 2008). At the end of the clashes, thirty-four civilians and twenty-seven pro-government and thirty-nine opposition activists were dead as well as a number of other casualties in the ranks of the police and the army (Daily Star).

It is not my aim here to make any political judgement on the outcome of the fighting. What is important to emphasize is that events on the ground were very different from international media representations of Lebanon's descent into chaos and of a 'takeover' by a terrorist group. Actors usually excluded from mainstream discourses of sovereignty enacted localized, effective control of territory and population through attempts to change specific components of the urban infrastructure, with complex effects for the everyday life of communities caught up in the conflict:

> The surprise element ... is that Hizbollah walked in largely unopposed, there were no rival militias to fight. Both Hariri and Joumblat issued instructions to their supporters not to enter into battle with Hizbollah or any other opposition forces, and to seek protection from the Lebanese army. Hizbollah, which has always justified itself as a resistance against occupation, became the occupier. (Shehadi 2008, 13)

In May 2008 Hezbollah became a hegemonic actor not simply by eroding the state's legitimate right to exert political violence but by producing new, hybrid forms of sovereignty and administration of the urban space. This was achieved not simply in opposition to the state but in close coordination with it.

Hybrid sovereignty is not the sole prerogative of Hezbollah. Government parties also used armed militias to maintain control of territory, and armed private security contractors are by no means politically neutral. At different moments both in the civil war and in the clashes of May 2008 the government, army, militias, private security contractors, and, indeed, foreign states cooperated and counteracted the sovereign practices of the others. These complex relationships dissolved the boundaries between the categories of state and nonstate. The urban environment was the terrain on which these practices were realized. Sovereignty appears here not as a fixed state but as hybrid, fluid, contested processes, knowledges and practices alongside, with and beyond the state.

Traditional approaches that see sovereignty as something which can be strengthened, measured and fulfilled prove inadequate for understanding political violence in Lebanon. An exclusive focus on Lebanon's weakness with respect to a vision of sovereignty as either state-led or absent overlooks a series of other de facto sovereignty arrangements – which I have called 'hybrid sovereignties'. This approach appreciates the role of nonstate actors not as anomalies, but as agents of sovereignty through their use and control of urban space. Here, hybrid political actors exist between the state and the nonstate; they are not the state but resemble it, collaborate with it, or overpower it. These hybrids are more than the sum of state and nonstate: they constitute new entities that are both state and nonstate, entities which enact a hybridized sovereignty born from this cross-contamination. These hybrid sovereignties are intimately linked with the management and control of the everyday urban built environment. The case of the May 2008 clashes is emblematic in this sense as territorial control and political violence were deployed through specific uses and targeting of particular urban spaces and infrastructures.

This last chapter serves to show how, rather than an exceptional 'tragedy' and descent into chaos, the Two Years' War instead is the urban manifestation of gradually hybridized sovereignty arrangements that continue to characterize moments of conflict in Lebanon nowadays. I have considered here a series of hybrid geographies where state and nonstate act in interconnected ways, and looked at how they do so amidst specific built environments. These hybrid geographies transcend the accepted binaries that mark realist views of sovereignty: state/nonstate, legitimate/illegitimate, order/chaos, national/urban, domestic/foreign. Furthermore, in contexts of conflict like Lebanon's, an urban vocabulary of sovereignty is needed to enable richer contextual understandings that allows to look beyond what sovereignty is supposed to be (de jure), and instead recognize how sovereignty *actually works* on the ground (de facto). In Lebanon and elsewhere, studying hybrid sovereignties makes it possible to interpret geographies of conflict beyond the logic of the state, and to account for spaces that are by no means chaotic, but where complex hybrid articulations of both state and nonstate become part of evolving everyday practices and material networks.

The built environment of Beirut is key to understanding the everyday practice of sovereignty beyond the perspective of traditional state and international relations theory. Both in 1975–1976 and in 2008, Beirut's built fabric became a contested ground for different views of territoriality, identity, and security, as framed by different non-state actors. In both instances, state and non-state actors, often through hybridized relationships, have framed, projected, and managed power and territory through interventions on urban infrastructure. This perspective poses fundamental questions about the nature of state sovereignty in Lebanon and indeed elsewhere, shifting from conventional notions of territorial sovereignty as the monopoly of the state, and towards more nuanced forms of exercise of political violence and territorial control that are instead practised by *both* state and non-state actors in more complex concertation.

# Conclusion

> *It is somewhat arbitrary to try to dissociate the effective practice of freedom by people, the practice of social relations, and the spatial distributions in which they find themselves. If they are separated, they become impossible to understand. Each can only be understood through the other. (Foucault 2000, 256)*

This book has explored the relationship between geopolitics, conflict, and built environment in Lebanon and its capital Beirut focusing particularly on the early but crucial phases of the 1975–1990 civil war. Chapters 5 and 6 counterposed official geopolitics and their rhetoric and policy decisions towards Lebanon with the unofficial militias' localized understandings of Lebanon's territory and role as a nation. Chapter 6, based on the urban theories outlined in Chapter 1, also showed how Beirut's built environment was formative in translating official geopolitics discourses into localized configurations of power and territory. Lebanon's spatial history of sectarianism, analysed in Chapter 2, has provided the context of analysis of the escalation towards the civil war (Chapters 3 and 4) and, most importantly, the modern colonial episteme in which both official geopolitics and militia urban territorialities are inscribed. Chapter 7 has shown how localized geopolitical understandings continue to re-emerge in more recent moments of intra-state conflict and how urban space and infrastructure are constitutive of hybrid sovereignty arrangements and practices that continue to develop beyond the state/nonstate divide. Thus, rethinking conventional sovereignty (as exercised exclusively by the State) to include the de-facto sovereignty practices of non-state actors, the notion of *hybrid sovereignties* in this book allows to step beyond discourses about Lebanon as a 'weak' or 'failed' state that tend to naturalize instability as chronic and inevitable, and instead propose thicker and context-based approaches to conflict.

From the exploration in this book, we can identify three core characteristics of Beirut's *urban geopolitics*:

1. State-based official geopolitics and nonstate militia narratives were not isolated and irreconcilable. Armed militia drew on official geopolitics to

shape and legitimize localized political and territorial projects. Official geopolitics chose a policy of non-intervention and ignored the urban dynamics of the conflict, which allowed militia territorialities to develop, crystallize and lay the bases of the partition of the city for the duration of the whole civil war.
2. Both official and militia geopolitics, and their understandings of Lebanese sovereignty and territoriality, operate within the same modern episteme that, during Lebanon's Ottoman and European imperial and colonial past, has at once produced the governmental dispositif of the sect and normalized it as age-old. While official geopolitics considered the war as dictated by unavoidable sectarian rivalries, the demise of the state during the Two Years' War was not replaced by 'chaos'. Instead, it opened the ground for a 'turbo-charged' political sectarianism that materialized in the militias' spatial and infrastructural reorganization of the city.
3. The built environment is part and parcel of these urban geopolitics. The urbicide of Beirut did not simply consist of violent and wanton targeting of architecture. Instead, it entailed a more complex reorganization of urban spatiality – the relation between people and things in the city – as part of localized interpretations and implementations of official geopolitical discourses.

In what follows, I outline these points more in detail, before envisioning future possibilities that this study offers for research and praxis around the contemporary urban geopolitics in Lebanon.

## De-subjugating geopolitics

In the last two decades, political geographers have developed critical interpretations of state weakness in postcolonial and (post)conflict situations (Agnew 2009; Lunstrum 2013; McConnell 2009; Mavroudi 2010; Ramadan 2009; Sidaway 2003; see also Mountz 2013). Albeit in very different ways, these all conceive sovereignty beyond the Westphalian system of state borders and account for sovereignty arrangements that do not follow those territorial geometries, and instead deploy within the spaces and infrastructures of the state.

The urban geopolitics outlined in this book is inspired by and seeks to contribute to this developing corpus of knowledge. It has adopted the notions of genealogy (Foucault 2003) and contrapuntal reading (Said 1994) to explore side by side the understandings of Lebanon's sovereignty and territory in the official geopolitical discourses of state governments *and* within by the militias involved in the urban battles. So what does an urban geopolitics of Lebanon tell us about the nature of the civil war that a state-scale geopolitics does not?

A counterpoint is not aimed at re-creating hierarchies, and an anti-geopolitical eye, as seen in the Introduction, ought to be critical without imposing further geopolitical truths. An urban geopolitics is neither 'by definition necessarily "better" than a national one [and] [u]rban geopolitical imaginations are not by definition any more democratic or pluralist than national ones' (Bialasiewicz 2015, 321). Rather than declaring the superiority of one set of representations over the other, the book has instead highlighted their constructed and contested nature, and their mutual relationship, by enacting 'an analytic that is contingent on context, place, and time, rather than a new theory of geopolitics or a new ordering of space' (Hyndman 2007, 36). This analysis disrupts accepted narratives, unsettles the scientific certainty of official geopolitics and de-subjugates differential and 'illegitimate' geopolitical knowledges from the urban battleground, thus 'creat[ing] the space in which different voices, dispositions, and behaviours might begin their own conversation' (Lisle 2016, 294).

By means of its contrapuntal analysis, this study aimed to read beyond the nation state centred perspective of foreign policy – the 'territorial trap' – and question instead the absences that those discourses implied. One example of these absences is the lack of detailed consideration of the militias and their geopolitical narratives and the impact of these differential narratives on the redefinition urban territoriality. This approach does not stop at deconstructing and highlighting the partiality of official geopolitics. Most importantly, it denounces the enormous urban impact and concrete consequences of foreign policy choices that were blind towards the urban scale and the socio-material processes – division, checkpoints, sniping, summary executions, purging of camps and towns, and partition – that were changing the political geography of the city.

'De-subjugating' should not be understood as simply revealing otherwise hidden and passive knowledges that operate in isolation. Instead, it has a more active meaning: it exposes those knowledges and (re-)attributes specific agencies to them. Militia knowledge and practices did not develop in isolation from established geopolitical discourses – such as Arab Nationalism and the post-colonial remainders of the Lebanese question – but drew on and renegotiated them locally, acquiring power and impacting the material layout and geographical imaginations of the city. These knowledges did not directly enter the maps of global politics, but they changed the daily life as well as the built fabric of Beirut.

## Decolonizing sectarianism

This book has adopted a post-colonial perspective on sectarianism that questions the actual existence of the sect as an immutable entity that pre-exists society and situates it within historically and geographically specific processes mechanisms of power. Sectarianism, as seen in Chapter 2, is a dispositif, a tool of governmentality. It is the political and cartographic result of Lebanon's complex encounter with European colonialism and the gradual changes and final demise of the Ottoman Empire over the course of the nineteenth and early twentieth centuries. Both official foreign policy accounts of the Two Years' War and its non-state narratives rely on this cartographic and political certainty. In Chapters 5 and 6, I have attempted to highlight how both sets of accounts are connected with modern colonial epistemologies that produced political sectarianism.

The 'sectarian nature' of the Two Years' War was a recurrent script in the geopolitical representation of Lebanon. The agency of Lebanese political leaders was reduced almost exclusively to an abstract sectarian inclination rather than contextual and tangible power arrangements. This sort of sectarian determinism, we have seen in Chapter 5, sanctioned even further an international policy of non-intervention while Beirut was partitioning, as it reinforced the view of a conflict that was intrinsic to the Lebanese social structure and thus unavoidable.

Colonial modernity produced not only new subjectivities (Kastrinou 2016; Makdisi 2000) and ways of envisioning territory before Lebanon's independence, but continued to reproduce its own taxonomies during the civil war. After an era of relative peace between the French mandate and the early 1970s, the sect as dispositif had become glaringly materialized in the militias' territorial partition and reorganization of Beirut. One of the legacies of Cartesian modern taxonomy has been that of creating orders of representation and having made these orders appear as 'a framework that seems to precede and exist apart from the actual individuals and objects ordered. The framework, appearing as something pre-existent, non-material and non-spatial, seems to constitute a separate, metaphysical realm – the realm of the "conceptual"' (Mitchell 1988, 176). The Lebanese sectarian order has become a normalized structure, supposedly existing a priori from the social process.

Lebanese sectarianism is often conceived as an un-modern phenomenon that originates from an undefined traditionalism or fanaticism. Within this vision, Lebanon has been often considered an 'unachieved state' (Salam 2001, 9). In fact, as we have seen, sectarianism has been part and parcel in the formation of the modern state and its territorial project: the sect was the unit around which every map of Lebanon produced since the 1840s was constructed. From this perspective, it is possible to see militia spatial practices during the civil war – ID killings, purging of villages and camps, population transfers, sniper killings, the erection of barricades, institution of enclaves – less as the sign of the disappearance into chaos of the modern state, and more as a post-colonial 'turbo-charging' of the sectarian dispositif.

Lebanese political and social life are based 'on a myth of communal homogeneity – that there is such a thing as a Maronite or a Druze nation that can or should be represented – and on a myth of traditional religious tolerance and harmony' (Makdisi 2000, 163). This myth idealizes the sect as a priori situation instead of recognizing it as sociocultural construct and creates a deadlock condemning Lebanese society to be chronically haunted by violence.

Such assumptions – which we saw were rife in the treatment of the Lebanese conflict in international foreign policy in 1975 and 1976 – can in turn be exploited by those perpetrating violence, in erasing their agency and responsibility in and for it.

Without dismissing the real, material and bloody effects that sectarianism has had on Lebanon since the 1840s at least, we need, however, to de-colonize the sectarian dispositive. We ought to conceive and interpret violence in its contextual specificity, 'not outside the realm of human society [or] as a devolution into a seething "proto-" or "pre-cultural" sets of behaviours' (Nordstrom and Robben 1995, 3) and be critical of disembodied, ahistorical views of violence as determined by unspecified 'ancient' rivalries.

## Contextualizing urbicide

As highlighted in Chapter 1, when considering 'the implosion of global and national politics into the urban' (Graham 2004b, 7), we risk to uncritically rescale conflict from something that used to happen at national borders to something that has become predominantly city-focused. This process risks, in turn, normalizing urban wars as a given symptom of our contemporary, post-Cold War moment (Sassen 2018, 2010). As a critique to this process, I have approached the arguments around urbanization of warfare from a widened historical and theoretical perspective.

First, part of this book addressed the links between political violence and urban built environment in a conflict preceding the post-cold war era – of which urbicide is often labelled as typical (Coward 2009). Second, it explored the built environment as a site of convergence of multiple and interconnected geopolitically relevant scales, 'from superpower geopolitics to city streets' (J. D. Sidaway et al. 2014, 1182). This way, it was possible to explore urbanized warfare not merely as a symptom of a specific geopolitical epoch, but rather, as the grounding of specific of contested geopolitical imaginations. Third, the book counters generalizing assumptions that cities have always been targets throughout history, and highlights the need to understand urbicide as a context-specific violent intervention on the spatiality of a city.

As seen in Chapter 6, during the Two Years' War attacks against Beirut's built environment were at once tactical *and* political. Urbicide was neither purely a mechanical act dictated by (para)military material necessity, nor simply a targeting of symbolic buildings for political reasons. Urbicide, instead, was a violent reorganization of the city's spatiality. Far from being an inert background where violent events unfold, Beirut's built environment is an active component of the violent reconfiguration of this spatiality. This reconfiguration can include drastic events like purging and razing refugee camps and villages for nationalist and sectarian reasons (Karantina, Damour and Tell Zatar), and substantial population displacements due to destruction of infrastructure and fear for safety. But it also encompasses more mundane dynamics like the reorganization of everyday infrastructure. Public sector and infrastructure jobs, allocated according to sectarian quotas, were disrupted by government and municipal employees and workers fleeing from specific areas into other neighbourhoods, which started the process of self-management of what were once state-owned services by the various militias in control of a given area (Harik 1994). As Harik sharply states, militia takeover of infrastructural and public services in those spaces that had been purged of the 'non-belonging' sect, had in fact as their ultimate goal political projects of 'community and national hegemony' (Harik 1994, 51). Urban infrastructure and built environment, in other words, had become geopolitical machines, actively employed to recast the national and geopolitical imaginations of specific communities and their urban space.

It is those urban logics and nonstate rationalities of the conflict – militia territorialities, but also the redesigning and re-managing of neighbourhoods and their infrastructure – that official diplomatic and foreign policy circles overlooked. Oversimplifying the conflict as chaotic and illogic, the reasonings of official geopolitics did not consider the contextual socio-material changes taking place in the city, attributing the violence instead to disembodied and unspecified 'fatal mechanisms' intrinsic to Lebanese society. This missed a crucial opportunity to redress the conflict by acting on its urban space.

## Towards pacific urban geopolitics

This book used the notion of urbicide to make sense of the impact of political violence on the socio-spatial arrangements that constitute the life of a city, and of which the built environment is a constitutive part. In this last part, I suggest that the built environment can be part and parcel of a socio-spatial reorganization towards durable peace. In what is portrayed as an era of urbanized and protracted warfare, we ought to ask whether and how urban spaces have a specific potential for fostering durable peace.

There is a recognized gap between the disciplines of geography and peace studies. Research in subdisciplines like political geography and critical geopolitics has, until recently, dealt more with war than with peace (Koopman 2011; McConnell, Megoran, and Williams 2014; Megoran 2011, 2010; Williams 2015). A 'pacific geopolitics', instead, should be on the agenda alongside mainstream approaches to geopolitics centred on space, politics and war (Megoran 2010). I want to develop this argument further and suggest that research on geographies of peace ought to connect more with architecture and planning literature around urban conflicts. In the last decade at least, this literature has studied how dynamics of conflict and peace are mutually shaped with the volumetric aspects of cities: from architecture and infrastructure (Brand and Fregonese 2013; Calame and Charlesworth 2009; Dumper and Pullan 2010; Pullan et al. 2007; Pullan 2011), to planning (S. A. Bollens 2006; S. Bollens 1999; Morrissey and Gaffikin 2006; Rokem and Boano 2017), to the everyday civic sphere (Pullan 2006), this literature has studied how cities absorb, deal with and *transform* conflict.

If, on the one hand, 'the multiplying relations between conflict and space have become one of the most pressing issues in geo-cultural research' (Mörtenböck and Mooshammer 2008, 75), on the other hand, however, the study of cities and geopolitics has not dedicated enough attention to the relations between *peace* and space. Urban geopolitics has mainly dealt with political violence in and militarization of cities and, more specifically and as seen in Chapter 1, with highly militarized contexts. According to Megoran (Megoran 2011, 178) if geography has to 'play a serious role in addressing the problems wracking twenty-first-century humanity, it is imperative that this imbalance be redressed'. This argument urgently needs to be applied to urban geopolitics.

By 2050, two third of the world's population will be urbanized and protracted armed conflict has also become a distinctively urban challenge: 'Wars have moved into the lives, cities and homes of ordinary people in a more vicious way than ever before' (ICRC 2017). This change in the geography of conflict is not only endangering the material fabric of cities worldwide; it is also causing unprecedented damage to urban communities, as nearly 50 million people (ICRC 2017) suffer the long-term impact of conflict in cities. If we are to take up what constitutes one of the most important twenty-first-century challenges, then urban geopolitics ought to account not only for the urban implications of war, but also for the urban conditions of peace: how peace is 'constructed and legitimated and implemented, how it gains consent, and how its actors learn' (Megoran 2011, 182). Equally, geographies of peace ought to engage more with the urban space and built fabric as a crucial arena for humanitarianism (Carpi and Boano 2018; Reid-Henry and Sending 2014), reconstruction and conflict transformation (Kurgan 2017; Vignal 2014).

How can urban space be an agent of pacification, how can the space of cities produce 'positive peace' (Galtung 1969; McConnell, Megoran, and Williams 2014) intended as a condition of enduring, structural harmony, rather than merely an absence of violence?

## Beirut's possible spaces for peace

In 2005, the killing of former PM Rafiq al Hariri and a following wave of targeted political killings dramatically impacted on Lebanon's political scene. What began was a period of mass protests – known as the Cedar Revolution – that apart from producing new political alliances and personalities, changed the meaning and use of Beirut's reconstructed city centre from a neoliberal privatized spatial design as a neutral space and home to the main state institutions, into a space for politics and activism, protest and encampment. Since then, new physical barriers (road closures, checkpoints, traffic diversions, and surveillance technologies), this time to protect the security of politicians both in the government buildings and around their private residences, altered the urban landscape, produced unequal mobilities (Brand and Fregonese 2013; M. Fawaz, Harb, and Gharbieh 2011; M. Fawaz, Harb, and Gharbieh 2012), and kept the city in a state of simmering tension, before it burst into the clashes of May 2008. After the election of a new President and the 2009 elections, relative stability returned, but in 2014, failing to agree on a new electoral law and to choose a new president until late in 2016, an era of impasse started.

And yet, during this era of apparent state-level inertia, a new constellation of social movements born in the aftermath of the 2005 protests began to give way to new types of political organization, campaigning away from the networks and spaces of the traditional parties (Battah 2018). New civil society movements – non-sectarian, diverse, activist-led, and rallying around urban and material causes – have started populating the political scene; their organizational power is strengthened, and their resonance amplified by the use of social media. From the *Laïque Pride* movement protesting sectarianism and in favour of civil marriage, to the *#YouStink* protest movement responding to government corruption and failure around refuse collection services which plunged the city into a massive garbage crisis in the Summer of 2015, such movements appear well suited to reinvent urban policies tailored to the particularities of the city. I argue that these new expressions also have the potential to turn into an urban politics that 'can counter the chauvinisms of national regimes' (Alsayyad and Roy 2006, 2).

Despite their lack of further development into established political programs, these early protest movements created a platform for more organized political expression. *Beirut Madinati* is an example of an anti-establishment politics born out of anti-government protest, taking over the space (literally and discursively) of what was perceived as a corrupted and inept government, attempting to renegotiate the societal and spatial mechanisms characterizing urban life, and at the same time being conscious of the legacy of urban conflict. Beirut Madinati 'seeks to build a political alternative from the local level outside sectarian frameworks' (beirutmadinati.com). Its demands for better urban infrastructure aim to 'strengthen the social fabric in one neighborhood and between the neighborhoods and consolidate the culture of the right to the city' (beirutmadinati.com) and ultimately improving urban knowledge among the public, from waste management to affordable housing, from pollution to green spaces, all of this often using inner city public spaces for electoral forums. After gaining a substantial result in the municipal elections in 2016 (the first since the Syrian crisis and the influx of refugees into the city) Beirut Madinati has become an established presence in the municipal political landscape, with one of its candidates elected head of the city's Order of Architects and Engineers in 2017, a crucial organization in influencing urban spatial politics.

7 May 2018, exactly ten years from the May 2008 clashes and sixty years from the assassination of Nasib Metni which ignited the first violent crisis since the country's independence, saw the first parliamentary elections Lebanon held in nine years. With a new electoral law, a generation of people between twenty-one and thirty years old was now voting for the first time. The translation of grassroots and municipal anti-establishment movements onto the national electoral arena was this time the list *Koullouna Watani*. Gathering eleven campaign groups, but also containing members coming from the traditional party system, Koullouna Watani aims at fighting corruption at incompetence in government. It is also using the city space differently to make political points, turning it into part and parcel of its message: from holding press conferences at rubbish dumps, to organized acts to subvert traditional parties' use of urban space for electoral campaigning, such as covering up campaign posters on walls with reflective paper to 'bringing the campaign focus back to the voters' (Battah 2018).

This book focused on the Two Years' War as a crucial moment of violence that has marked Beirut's urban spatial politics for decades, and still does. The spatial issues highlighted by the politics of Beirut Madinati and Kulluna Watani become even more crucial questions for a city like Beirut, where many of the answers about the civil war that are necessary for creating durable peace, remain unresolved. Lebanon's 1991 amnesty law led to the absence of state 'truth and reconciliation' enquiries into war responsibilities and crimes. A state-led discourse about memory and reconciliation and a collective narrative around the causes and events in the civil war is missing. The urban context surrounding this absence is one of spatial, infrastructural and mobility inequalities (many pre-dating the civil war) that Beirut's neoliberal reconstruction has contributed to reinforce (Khalaf and Khoury 1993; S. Makdisi 1997; Monroe 2016; Nagel 2002; Sawalha 1998, 2000). This situation has left individuals and civil society to produce and curate a multipolar historical culture 'in which individual memories negotiate with state-led amnesia, and trauma-induced absences of discursive space coexist with civil-led organisations determined to put recent community history on record' (Launchbury 2014, 105). Currently, important practices and sites where memorialization is attempted are present in Lebanese civil society. These include: localized and intergenerational memory cultures (Haugbølle 2010; Larkin 2012); private archives (Launchbury 2014); performance and oral history initiatives around memory and reconciliation (Rowayheb and Ouaiss 2015 See also: fightersforpeace.org); and grassroot-sourced cartography of the civil war missing people, that maps the location of former burial places, interrogation rooms, and disappearance spots, in the eventuality that future governments decide to trace responsibility for past violence (www.actforthedisappered.com). It is in these sites that 'the understudied republic that is the infrastructure of the modern city can become the main focus of political action' (Amin and Thrift 2017, 6) and a pacific urban geopolitics be enacted.

This book never aimed to provide a systematic chronicle or comprehensive historical account of Lebanon's Two Years' War. Neither does it make conclusive claims about the country's darkest moments of violence or the complexity of its sovereignty arrangements. Most importantly, it does not intend to lecture the people of Lebanon on the causes, course and possible futures of the political violence that they, in so many forms and intensities, have experienced and still experience.

However, these pages have shown how Beirut is a geopolitical city, where multiple scales of political violence become grounded in urban space and infrastructure, and where change towards peace can and must start from accounting for the built environment and for the residents' everyday experiences of it, rather than from traditional national rhetoric. Urban space and infrastructure thus become political conductors, they are part of the production of political knowledge and action beyond the state. Furthermore, it is not only the built environment and infrastructures that are 'switched into being as a political force' (Amin and Thrift 2017, 6), but the complex, political and understudied physical elements that interact with and crucially shape the city, such as air, electricity, concrete, soil. Engaging forensically with how these elements shape urban politics in light of the long-term impact of war can open the field of enquiry to an elemental urban geopolitics oriented towards durable peace.

# Notes

## List of participants from the 1975–1976 war

1   *Film Ameriki Tawil* (a long American film) is a 1980 theatre play by Lebanese musician and playwriter Ziad Rahbani. The play is set in a psychiatric hospital in West Beirut during the civil war. Here, the play follows the individual stories of the patients, whose words seem to make more sense than those of the medics in charge.

## Introduction

1   Although the Document of National Reconciliation (Taif Agreement) was signed in Taif (Saudi Arabia) on 22 October 1989, violence continued on the ground through 1990 – with the assassination of neo-elected president elect René Muawad and fighting between General Michel Aoun, the Lebanese Forces and the Syrian army – until the formation of a government of national reconciliation on 24 December 1990. In 1991, the disbandment of all militias (bar Hezbollah and the South Lebanon Army), the treaty of cooperation between Lebanon and Syria, the re-establishment of control by the Lebanese army over the PLO and the granting of amnesty for all crimes committed during the civil war (26 August), and Aoun's exile to France are major events that de facto mark the halt of violence and the re-establishment of basic institutional functions in the country.

2   On 14 March 1978, the Israeli army entered south Lebanon initiating Operation Litani, a heavy artillery and naval bombardment campaign of occupation in retaliation to what became known as the coastal road massacre three days earlier, when a commando of Palestine Liberation Organization fighter killed thirty-eight Israeli civilians after hijacking two buses on the road between Haifa and Tel Aviv.

3   At the end of the Two Years' War and throughout 1977, a truce held despite a number of skirmishes and harbingers of further complications. The south of Lebanon was in turmoil, though, as Israel and the Palestinian resistance often confronted each other; the Israeli army used air power against the Palestinian resistance in the Lebanese villages; moreover postwar economic recession doomed the country. While peace slowly returned, diplomatic visits followed each other and international aid arrived. Nevertheless, tensions ran high as various political personalities lost or risked their lives in car bomb attacks and other types of sabotage. On 14 March 1978, Israel launched 'Operation Litani'. The operation aimed at creating a buffer

zone free from the Palestinian resistance from the Israeli northern border to the banks of the river Litani. The events attracted the attention of the international community with wider resonance than in the past phase: the United Nations passed resolution 425 and 426 calling for the respect of the territorial integrity of Lebanon and instituted the United Nations Interposition Force in Lebanon (UNIFIL). The Israeli Defence Force (IDF) retreated in the same month, leaving the control of the buffer zone to the South Lebanon Army headed by Christian major Saad Haddad. In June 1982, following the assassination of the Israeli ambassador in London by the group Abu Nidal, the IDF began 'Operation Peace for Galilee' during which air strikes were carried out regularly on Lebanon and especially against the city of Beirut.

4   I adopt here Michel Foucault's definition of 'power/knowledge' as the terrain in which webs of power and visions of what constitutes reality mutually interact: 'Perhaps [...] we should abandon a whole tradition that allows us to imagine that knowledge can exist only where the power relations are suspended and that knowledge can develop only outside its injunctions, its demands and its interests [...] We should admit rather that power produces knowledge (and not simply by encouraging it because it serves power or by applying it because it is useful); that power and knowledge directly imply one another; that there is no power relation without the correlative constitution of a field of knowledge, nor any knowledge that does not presuppose and constitute at the same time power relations. [...] [I]t is not the activity of the subject of knowledge that produces a corpus of knowledge, useful or resistant to power, but power-knowledge, the processes and struggles that traverse it and of which it is made up, that determines the forms and possible domains of knowledge' (1997, 27–28).

5   These are obviously not the only countries whose voice was influential in the international foreign policy towards Lebanon at the time, and it would be important to include how other states scripted Lebanon geopolitically, for example, considering the perspective of Russia on the conflict and on potential escalation in the Middle East. The perspective of other Arab countries – such as Syria and Egypt – whose representatives were often involved in diplomatic activity with Lebanon during the war, and eventually became involved militarily through the entrance of the Arab Deterrent Force (ADF), should also be considered.

# Chapter 1

1   Examples of asymmetric conflicts include: that between the Tamil Tigers and the Sri Lankan army concluded in 2009; between the Israeli army and Hezbollah in Lebanon in summer 2006; between the armies of the coalition

and Iraq's insurgencies since 2003; between Islamic State and regular armies across Iraq and Syria; between the Syrian government (with the foreign armies engaged in Syria) and the various local resistances as well as irregular armed groups like Hezbollah fighting in support of the Syrian government; and in Yemen between Houthi insurgencies and the Yemeni government.

2   Convention (IV) respecting the Laws and Customs of War on Land and its annex: Regulations concerning the Laws and Customs of War on Land. The Hague, 18 October 1907.

3   In 1987, the number of religious buildings and places of worship in Lebanon damaged by the violence of the civil war was 437 across all confessional communities (Labaki and Abou Rjeily 1993, 162). Of these, 57.3 per cent places of worship were in the suburban and rural areas of the Mount Lebanon Governatorate, precisely in the districts of Aley, Babda and Chouf (250 Christian; 41 Sunni Muslim). The worst damage to Shiite places of worship happened in the Governatorate of South Lebanon (17). Only 11 per cent of the damages to religious buildings were in Beirut (Labaki and Abou Rjeily 1993, 164–69).

# Chapter 2

1   The noun tā'ifa (religious community, sect) originates from the adjective tā'ifiyy (multi-communal/communal). Palestinian scholar Anis Sayigh employs a twofold use of the concept: on one hand, he intends 'multi-communal society'; on the other hand, he uses with the meaning of 'confessional society', i.e. a society whose multi-communal features have been politicized by interested actors that Sayigh calls al- tā'fiyyūn (for more details, see Firro 2003, 49).

2   The term *dispositif* (or apparatus) is used here as defined by French philosopher Michel Foucault, as 'a thoroughly heterogeneous ensemble consisting of discourses, institutions, architectural forms, regulatory decisions, laws, administrative measures, scientific statements, philosophical, moral and philanthropic propositions - in short, the said as much as the unsaid' (Foucault and Gordon 1980, 194). In our case, this definition is useful because the sect as a modern project became embedded, starting in the nineteenth century, in a multifarious universe of discourses and practices of subjectification and identification – a new episteme – in Mount Lebanon, as illustrated by Makdisi. Moreover, Foucault understands the strategic function of the dispositif as a responsive process to an 'urgent need' in a precise historical moment (194–95). As such, the dispositif is intimately related to power shifts and new developing power relations – as was the case in the transition from the feudal power structure of the Emirate to the modern and territorial power distribution based on the

sect: 'it is a matter of a certain manipulation of relations of forces, either developing them in a particular direction, blocking them, stabilising them, utilising them, etc. The apparatus is thus always inscribed in a play of power, but it is also always linked to certain coordinates of knowledge which issue from it but, to an equal degree, condition it. This is what the apparatus consists in: strategies of relations of forces supporting, and supported by, types of knowledge' (196).

3   Writing about romantic travel writing, Thompson (2012) illustrates several examples of how colonial ventures were interlinked with the cultural world of the *voyages pittoresques* and how exotic writing often resulted from close proximity to military events and violence. Theophile Gautier accompanied as an observer governor-general Thomas Robert Bugeaud to Kabylia to quell local revolts. Dominique-Vivant Denon, himself invited to join Napoleon's expedition to Egypt and subsequently appointed first director of the Louvre, declared during his visit to Egypt 'si l'amour de l'antiquité a fait souvent de moi un soldat, la complaisance des soldats pour mes recherches en a fait souvent des antiquaires' (Thompson 2012, 21).

4   In her chapter on Beirut's sectarian relations from the mid-nineteenth century, Tarazi-Fawaz (1983, 116–17) gives an extended account based on primary evidence of escalating episodes of violence and murder between the Muslim and the Christian communities in the city. She also points out other ways of enacting sectarian contested identities, especially among the urban rich: the creation of schools (such as the Christian Maronite *Collège de la Sagesse* created in 1874), of hospitals and charitable foundations (for example, the Sunni Muslim Maqassed benevolent society funded in 1878), and of cultural and social institutions, many of which still operate in Beirut.

5   The *Convention between Great Britain, Austria, Prussia, Russia and Turkey for the Pacification of the Levant* was ratified in London on 15 July 1840. It proposed the boundaries of southern Syria beyond which Ibrahim Pasha had to withdraw and be granted hereditary sub-reign, leading to the pacification of the area: 'this line, beginning at Cape Ras el Nakhoura [...] shall extend direct from thence as far as the mouth of the River Seizaban, at the northern extremity of the lake of Tiberias; it shall pass along the western shore of that Lake it shall follow the right bank of the Rive Jordan and the western shore of the Dead Sea; from thence it shall extend straight to the Red Sear, which it shall strike at the northern point of the Gulf of Akab; and from thence it shall follow the western shore of the Gulf of Aqaba, and the eastern shore of the Gulf of Suez, as far as Suez' (Convention and Article. Pacification of the Levant [Also known as the Treaty of London]. The National Archives of the UK (TNA), FO 93/110/4A, 15 July 1840. See also Hertslet 1875, 1013).

6   The idea of Ottoman administrative and territorial decline is critiqued in Çırakman's (1996) account of the contradictory rather than monolithic European cultural imaginary of the Ottoman Empire, which depicted

the latter at once as hostile and weak (the sick man of Europe). Tarazi-Fawaz also shows the partial construction of the idea of Ottoman decline and contextualizes it within a narrative of decentralization. She traces back the idea of decline as contrasting with accepted European notions of sovereignty as centralized territorial power distributed uniformly within a bounded territory, while 'the Ottoman empire, far from being an "Oriental despotism" aspired for most of its history to a loose and subtle form of suzerainty over far-flung territories' (Tarazi-Fawaz 1994, 2). While, she argues 'the Eighteenth century saw the gradual loss of real power from the centre to the regions and accelerated processes of social class formation within regions. The weakening of the power of the Ottoman [...] did not entail, however, a more general social and economic decay. Decentralisation, rather than decline seems to capture better the broad historical trends of the Middle East [...] in that period' (Tarazi-Fawaz 1994: 2).

7   Wyld, J. (1840). Map of Syria- ancient and modern. 8 miles to one inch. J. Wyld, London.
8   After the fall of the Shehab emirate a number of Druze notables, exiled during the Egyptian-Ottoman war, had returned to reclaim their lands from the new ruler installed by the Ottomans, Bashir Qasim, who in the meantime had redistributed them. Thus, a dispute over land sparked first skirmishes between Christians from Dei al Qamar and Druzes from the nearby town of Baaqleen, followed by a three-week siege of Deir al Qamar (U. Makdisi 2000).
9   In particular, see pages 99–100 and Chapter 7.
10  This was formed by commissioners representing Britain (Lord Dufferin), France (Leon Béclard), Austria (De Weckbecker) Prussia (De Reyfus), and Russia (Nowikov). See (Hakim 2013, n. 21 Chapter 3, 284).
11  Reglement fondamental relatif à l'administration du Mont-Liban en date du 30 zilkadeh 1227 H; 9 Juin 1861. In Tarazi-Fawaz (1994, 216).
12  The special districts of Beirut, Jerusalem, and Samsun were also Mustasarrifiyyas.
13  This was a secret agreement reached on 15 and 16 May 1916 between French and British diplomats Francois Picot and Mark Sykes about the delineation of the territories of direct or indirect influence on the Eastern Mediterranean and its hinterland. The outcome was a map of a partitioned Middle East with an overview of the French and British interests – from customs fees to schools and hospitals, to railways, to government – within the bigger picture of the formation of an Arab state or a confederation of individual states, whose borders were to be traced on the matrix of these wider areas of influence.
14  The League of Nations report on the mandates system also observes, regarding Lebanon, the difficulty to capture precise demographic data, due to the diversity and mobility of the population: '[o]n taking up the mandate, the mandatory Power was confronted with a situation in which,

as regards demographic data, it had to start from the very beginning and under peculiarly unfavourable conditions: the population was suspicious, was made up of dissimilar races and belonged to antagonistic creeds; there were political and religious disturbances, vast areas to control, a part of the population was not settled (nomads and semi-nomads), etc. The task was, however, undertaken and a good deal of progress made, though, in 1938, no precise statistics for the country as a whole were yet available.' (League of Nations 1945).
15 The first was Decree N°60 L.R. in 1936. The Druze, Sunni and Sh'is communities opposed the organic law, so a second decree (N°146 L.R. in 1938) reversed their status, until all three eventually adopted the status of historic communities, gaining rights to be represented as separate communities within state institutions (See Firro 2006, 743).
16 Estimates by IRFED (then the Institut Internatioal de Researche et de Formation Education et Developpement), however, state 160000 (Ruppert and Verdeil 1999).

## Chapter 3

1 Between 1958 and 1961, Syria and Egypt joined their territories forming the UAR.
2 In 1952, a group of opposition politicians and prominent notables (the Social National Front, SNF), strong of wide swathes of public opinion, forced the President of the Republic Bishara al Khoury to resign after calling a general strike on 11 September 1952. Despite being analysed as a near-coup that signalled Lebanon's lack of strong leadership (Hudson 1985), the transition was sustained by the available state institutions (the Chamber of Deputies elected a new president, Camille Shamun), the opposition activity proceeded orderly and without violence, and the crisis left the political system intact (El-Khazen 2000).
3 The Eisenhower Doctrine was pronounced by then US President Dwight D. Eisenhower in his *Special Message to the Congress on the Situation in the Middle East* of 5 January 1957. It authorized the United States 'to cooperate with and assist any nation or group of nations in the general area of the Middle East in the development of economic strength dedicated to the maintenance of national independence. It would, in the second place, authorize the Executive to undertake in the same region programs of military assistance and cooperation with any nation or group of nations which desires such aid. It would, in the third place, authorize such assistance and cooperation to include the employment of the armed forces of the United States to secure and protect the territorial integrity and political independence of such nations, requesting such aid, against overt armed aggression from any nation controlled by

International Communism' (http://www.presidency.ucsb.edu/ws/index.php?pid=11007&st=&st1=). In line with the wider strategy of containment, the Doctrine involved those eastern Mediterranean states, like Lebanon, of high interest to US oil imports and that the United States considered 'on the edge' of falling into the Soviet or Arab Nationalist sphere of influence.
4   By which the United States and the Western bloc would provide financial and economic aid to Greece and Turkey to prevent them from falling into the Communist sphere. The text of the doctrine quotes: 'It is necessary only to glance at a map to realize that the survival and integrity of the Greek nation are of grave importance in a much wider situation. If Greece should fall under the control of an armed minority, the effect upon its neighbour, Turkey, would be immediate and serious. Confusion and disorder might well spread throughout the entire Middle East' (http://www.trumanlibrary.org/publicpapers/index.php?pid=2189&st=&st1=).
5   http://www.history.army.mil/documents/AbnOps/TABD.htm.

# Chapter 4

1   As seen in Chapter 1, Derek Gregory (2004) wrote of architectures of enmity as the ensemble of mappings rhetoric that underpin the policies and actions (including military) composing global foreign policy.
2   These estimates are based on Tables 5 (p. 29) and 9 (p. 36) in Labaki and Abou Rjeily (1993).
3   Agriculture figured as 18.5 per cent of the GDP in 1961 (under Fouad Chehab), decreased to 11.9 per cent of the GDP in 1964 (the start of Charles Helou's term), and continued to diminish until 9.33 per cent in 1974 (Labaki and Abou Rjeily 1993, 177).
4   Translated from the Arabic by Imad Aoun.
5   This paragraph has been reproduced from Ramadan, A., Fregonese, S. 2017. 'Hybrid Sovereignty and the State of Exception in the Palestinian Refugee Camps in Lebanon.' *Annals of the American Association of Geographers* 107, 949–63. Reprinted by permission of the publisher (Taylor & Francis Ltd, http://www.tandfonline.com).
6   This paragraph has been reproduced from Ramadan, A., Fregonese, S. 2017. 'Hybrid Sovereignty and the State of Exception in the Palestinian Refugee Camps in Lebanon.' *Annals of the American Association of Geographers* 107, 949–63. Reprinted by permission of the publisher (Taylor & Francis Ltd, http://www.tandfonline.com).
7   Differently from Hanf, Kassir identifies the birth of *Al-Tanzim* (which he describes as a new militia) in 1973, established by then mid-ranking officer and future general (and current President of the Republic) Michel Aoun (Kassir 2010, 508).
8   One of Pierre Gemayel's bodyguards killed on 13 April 1975.

9   The term 'isolationism' (*'azl; Inti'zal*) – used besides the term 'rejectionist' – identified the Kata'ib party and its allies (mainly the Ahrar party) after Kamal Joumblatt's call for it to be banned in the Spring of 1975 and after a common declaration by '23 pan-Arab and leftist parties and groupings drawn from several Arab countries and headed by Joumblat' and strongly influenced by Lebanese Communist Party leader George Hawi (el Khazen 2000, 288) announcing a plan for the political and economic isolation of the Kata'ib party on the Arab scene. Initially conceived as a measure to defuse tension and avoid military escalation, the isolationism policy had the result to instigate a highly defensive spirit among the Kata'ib ranks and Christian population, which began seeing the party as the only possibility against PLO military domination. The most eloquent effect of the isolationist policy consisted of Prime Minister Rashid el Solh presenting his resignation speech in the form of the accusation of the Kata'ib party of provoking the shooting of Ayn al Rummana, which signed the moment of the political wreckage when 'as he finished his […] speech, Solh left the podium and ran quickly outside the hall amidst the objections and shouting of many deputies. Kata'ib deputy Amin Gemayel hurled himself to the door trying to prevent Solh from leaving before hearing the Kata'ib party reply to his accusations' (el-Khazen 2000, 289. For photographic evidence of the moment, see also Chami 2005, 28).

# Chapter 5

1   FCO 93/678 (H.M. diplomatic service, dept Near East and North Africa) 'Internal political situation in Lebanon'. Paul Wright retired from diplomatic service in April 1975 and subsequently became chairman of the Anglo-Lebanese society.
2   The countries considered in the analysis for this chapter did not intervene with any major action in Lebanon prior to 1978, when Israel launched Operation Litani, and 1982, with the arrival of the Multi-National Force in Lebanon (MNF), the MNF was a four-nation peacekeeping force consisting of US, British, French and Italian army contingents. Its mission was also to train parts of the Lebanese Armed forces.
3   Fatahland was the name used by Israel to designate the mountainous Arqub area in that was under the control of the Palestine Liberation Organization (PLO) and especially the party funded by Yasser Arafat's *Fatah* (hence the name Fatahland) since the 1960s (see Chapter 4).
4   Jean Sauvagnargues was French Foreign Minister under President Valery Giscard d'Estaing from 1974 to 1976. He was succeeded by Louis de Guiringaud.
5   A critical view of the contested relationships between militia formation and sectarian identity is found in Rowayheb's (2006) study of cross-religious

militias and their mechanisms of construction of identity. Sectarian identity as a construct that is reproduced through context-specific interactions between militias and militia members, such as conscription, organization, membership and recruit. These practices explain the existence of various types of militias: if a number of them were recruited mainly according to religion, several of them were heterogeneous from the point of view of religious identity and drew on other parameters to construct their agendas, such as Arabism.

## Chapter 6

1. Translated by the author.
2. Neighbourhoods in what had become the western sector of the city.
3. Al-Mourabitun (Harakat al-Nassiriyin al-Mustaqillin), in Arabic meaning 'the sentinels' is the armed militia of the Nasserite Independent Organisation funded in 1957.
4. The Nazi-fascist inspiration of the Lebanese phalanges is tracked back to Pierre Gemayel attending the 1936 Berlin Olympic games as President of the Lebanese football federation, and praised the discipline in German society in the early years of the Third Reich, but without ever naming Hitler, and underlining that the word Nazism at the time did not mean the atrocities that we nowadays know. He pledged to export that discipline to Lebanon and upon his return he created the youth movement of the Lebanese Phalanges (Kataib).

## Chapter 7

1. IR scholar Gokham Bacik (2008) also uses the phrase 'hybrid sovereignty'. For Bacik, however, hybrid sovereignty is 'the product of the clash between the *de jure* and the *de facto* practices deployed in the colonially imposed state structures' (15) in postcolonial Arab political systems. The result is a 'hybrid sovereignty' which is 'neither Western nor traditional' (33), generating what are perceived as 'failures' of the Arab political system. Differently from its use in this book, Bacik's use of the term 'hybridity' reproduces normative views of sovereignty as a) a situation that can be accomplished and fulfilled and b) something whose accomplishment or not has a moral connotation of success vs failure.
2. This and the following two sections have been published by the author in *Environment and Planning D: Society and Space*, Volume 30, Issue 4, 2012, published by Sage Publishing, All rights reserved.

# References

Abourahme, Nasser. 2015. 'Assembling and Spilling-Over: Towards an "Ethnography of Cement" in a Palestinian Refugee Camp: Ethnography of Cement in a Palestinian Refugee Camp.' *International Journal of Urban and Regional Research* 39 (2): 200–17.

Abujidi, Nurhan. 2014. *Urbicide in Palestine: Spaces of Oppression and Resilience.* Routledge Studies in Middle Eastern Politics 63. London: Routledge.

Adams, Nicholas. 1993. 'Architecture as the Target'. *Journal of the Society of Architectural Historians* 52 (4): 389–90. https://doi.org/10.2307/990864.

Adey, Peter. 2013. 'Securing the Volume/Volumen: Comments on Stuart Elden's Plenary Paper "Secure the Volume"'. *Political Geography* 34 (May): 52–54.

Agnew, John. 1994a. 'The Territorial Trap: The Geographical Assumptions of International Relations Theory'. *Review of International Political Economy* 1 (1): 53–80.

Agnew, John. 1994b. 'The Territorial Trap: The Geographical Assumptions of International Relations Theory'. *Review of International Political Economy* 1 (1): 53–80.

Agnew, John. 1995a. *Mastering Space: Hegemony, Territory and International Political Economy.* London; New York: Routledge.

Agnew, John. 1995b. *Mastering Space: Hegemony, Territory and International Political Economy.* London; New York: Routledge.

Agnew, John. 2005. 'Sovereignty Regimes: Territoriality and State Authority in Contemporary World Politics'. *Annals of the Association of American Geographers* 95 (2): 437–61.

Agnew, John. 2009. *Globalization and Sovereignty.* Globalization. Lanham: Rowman & Littlefield Publishers.

Agwani, M. 1965. *The Lebanese Crisis, 1958 a Documentary Study.* New York: Asia Publishing House.

Aibar, Eduardo, and Wiebe E. Bijker. 1997. 'Constructing a City: The Cerdà Plan for the Extension of Barcelona'. *Science, Technology, & Human Values* 22 (1): 3–30.

Al Hawadess. 1975. 'Saint Charles. The Pyramid of Beirut', 28 February 1975.

Al-Harithy, Howayda. 2010. *Lessons in Post-War Reconstruction: Case Studies from Lebanon in the Aftermath of the 2006 War.* Planning, History and Environment Series. London; New York: Routledge.

Al Jazeera English. 2008. 'Nasrallah Addresses Lebanon - 08 May 08'. Al Jazeera English. http://www.youtube.com/watch?v=YrfxnCoPViI.

Alin, Erika G. 1994. *The United States and the 1958 Lebanon Crisis: American Intervention in the Middle East.* Lanham: University Press of America.

Al-Lawzy, S. 1975. 'Harb 'Al-Miraya' Bayna Al-Shiyyah Wa 'Ain Al-Rummana! [The 'War of Mirrors' between Al-Shiyyah and Ain-Al-Rummana]'. *Al-Hawadess*, 1975.

Al-Nahar. 1976. 'Saqata Al-Holiday Inn [Holiday Inn Has Fallen]', 22 March 1976.

Alsayyad, Nezar, and Ananya Roy. 2006. 'Medieval Modernity: On Citizenship and Urbanism in a Global Era'. *Space and Polity* 10 (1): 1–20.

Amar, P. 2009. 'Operation Princess in Rio de Janeiro: Policing "Sex Trafficking," Strengthening Worker Citizenship, and the Urban Geopolitics of Security in Brazil'. *Security Dialogue* 40 (4–5): 513–41. https://doi.org/10.1177/0967010609343300.

Amin, Ash, and N. J. Thrift. 2017. *Seeing Like a City*. Cambridge, UK; Malden, MA: Polity.

Anderson, Benedict. 1991. *Imagined Communities: Reflections on the Origin and Spread of Nationalism*. London: Verso.

Ashcroft, Bill, Gareth Griffiths, and Helen Tiffin, eds. 1995. *The Post-Colonial Studies Reader*. London; New York: Routledge.

Atkinson, David, ed. 2005. *Cultural Geography: A Critical Dictionary of Key Concepts*. International Library of Human Geography 3. London; New York: I.B. Tauris; In the United States distributed by Palgrave Macmillan.

Ballinger, Pamela. 2003. *History in Exile: Memory and Identity at the Borders of the Balkans*. Princeton, NJ: Princeton Univ. Press.

Barak, Oren. 2009. *The Lebanese Army: A National Institution in a Divided Society*. Albany: State University of New York Press. http://public.eblib.com/choice/publicfullrecord.aspx?p=3408025.

Battah, Habib. 2018. 'The Changing Culture of Election Campaigns in Lebanon'. *Beirut Report* (blog). 4 May 2018. http://www.beirutreport.com/2018/05/the-changing-culture-of-election-campaigns-in-lebanon.html.

Baudouı¨, R., and A. Grichting. 2005. 'Actes Du Colloque: Urbicide, Urgence, Durabilite´: Reconstruction et Me´moire'. In *Actes Du Colloque: Urbicide, Urgence, Durabilite´: Reconstruction et Me´moire*. Geneva: Institut d'Architecture de l'Universite´ de Gene`ve.

Berman, Marshall. 1984. 'Roots, Ruins, Renewals: City Life after Urbicide.' In *Village Voice*, 4 September, p. 20.

Berman, Marshall. 1996. 'Falling Towers: City Life after Urbicide.' In Geography and Identity: Exploring and Living Geopolitics of Identity, edited by D Crow, 172–92. Washington, D.C: Maisonneuve.

Bevan, Robert. 2005. *The Destruction of Memory: Architecture at War*. London: Reaktion.

Bhabha, Homi K. 2004. *The Location of Culture*. Routledge Classics. London; New York: Routledge.

Bialasiewicz, Luiza. 2015. 'An Urban Geopolitics'. In *Urban Europe. Fifty Tales of the City*, edited by Virginie Mamadouh and Anne, 317–22. Amsterdam: Amsterdam University Press.

Biersteker, Thomas J., and Cynthia Weber. 1996. 'The Social Construction of State Sovereignty'. In *State Sovereignty as Social Construct*, edited by Thomas J. Biersteker and Cynthia Weber, 1–21. Cambridge: Cambridge University Press.

Bishop, Ryan, and Gregory Clancey. 2004. 'The City-as-Target, or Perpetuation and Death'. In *Cities, War and Terrorism. Towards an Urban Geopolitics*, edited by Stephen Graham, 54–74. London: Blackwell.

Bogdanovic, Bogdan. 1994. 'The City and Death'. In *Storm 6: Out of Yugoslavia*, 53–57. London: Storm/Carcanet.

Bollens, Scott. 1999. *Urban Peace Building in Divided Societies – Belfast and Johannesburg*. Boulder, Colorado: Westview Press.

Bollens, Scott A. 2006. 'Urban Planning and Peace Building'. *Progress in Planning* 66 (2): 67–139.

Bou Akar, Hiba. 2005. 'Displacement, Politics, and Governance: Access to Low-Income Housing in a Beirut Suburb'. Cambridge, MA: Department of Urban Studies and Planning, MIT.

Bou Akar, Hiba. 2018. *For the War yet to Come: Planning Beirut's Frontiers*. Stanford, CA: Stanford University Press.

Boudjikanian-Keuroghlian, A. 1994. 'Beyrouth 1920–1991: D'une Métropole de Croissance Au Champ de Guerre'. In *Le Liban d'aujourd'hui*, edited by F. Kiwan, 241–73. Paris: CNRS Éditions.

Bougarel, Xavier. 1999. 'Yugoslav Wars: The "Revenge of the Countryside" between Sociological Reality and Nationalist Myth'. *East European Quarterly* 33 (2): 157–75.

Bourgey, A. 1985. 'La Guerre et Ses Conséquences Géographiques Au Liban'. *Annales de Géographie* 521 (January–February): 1–37.

Brand, Ralf, and Sara Fregonese. 2013. *The Radicals' City: Urban Environment, Polarisation, Cohesion*. Design and the Built Environment. Farnham: Ashgate.

British Embassy Beirut Report. 1975. 'British Embassy Beirut Report'. FCO 93/676. National Archives, Kew.

'British Embassy Report, 5 July'. 1975. FCO 93/676. 123–286.

'British Embassy Report, 19 April'. 1975. FCO 93/675. 1–122.

Brynen, Rex. 1989. 'PLO Policy in Lebanon: Legacies and Lessons.' *Journal of Palestine Studies* 18 (2): 48–70.

Brzoska, Michael. 2004. '"New Wars" Discourse in Germany'. *Journal of Peace Research* 41 (1): 107–17.

Calame, Jon, and Esther Ruth Charlesworth. 2009. *Divided Cities : Belfast, Beirut, Jerusalem, Mostar, and Nicosia*. Philadelphia: University of Pennsylvania Press.

Campbell, David. 1998. *Writing Security: United States Foreign Policy and the Politics of Identity*. Rev. ed. Minneapolis: University of Minnesota Press.

Campbell, David, Stephen Graham, and Daniel Bertrand Monk. 2007. "Introduction to Urbicide: The Killing of Cities? Theory and Event" 10 (2).

Carpi, Estella, and Camillo Boano. 2018. 'Border Towns: Humanitarian Assistance in Peri-Urban Areas'. *Humanitarian Exchange Magazine*. 2018.

Çelik, Zeynep. 1997. *Urban Forms and Colonial Confrontations: Algiers under French Rule*. Berkeley: University of California Press.
Chakhtoura, Maria. 2005. *La Guerre Des Graffiti. Liban 1975–1977*. Beirut: Dar an-Nahar.
Chami, J. (2005) *Chronicle of a war*. Beirut: mémorial du Liban.
Charon, Cyrille. 1905. 'La Syrie de 1516 à 1855 (fin)'. *Échos d'Orient* 8 (55): 334–43.
Chaslin, François. 1997. *Une Haine Monumentale: Essai Sur La Destruction Des Villes En Ex-Yougoslavie*. Paris: Descartes & Cie.
Clapham, Christopher. 1998. 'Degrees of Statehood'. *Review of International Studies* 24: 143–57.
Coaffee, Jon. 2003. *Terrorism, Risk, and the City: The Making of a Contemporary Urban Landscape*. Aldershot; Burlington, VT: Ashgate.
Cochrane, Paul. 2008. 'Lebanon's Media Battle'. *Arab Media and Society*, 29 September.
Command of Her Majesty. 1845. 'Correspondence Relative to the Affairs of Syria: Presented to Both Houses of Parliament'.
Corm, G. 2005. *Le Liban Contemporain: Histoire et Société*. Paris: La Découverte.
Corriere della Sera. 2008. 'Scontri e Morti a Beirut: Libano Sull'orlo Del Caos [Clashes and Dead in Beirut: Lebanon on the Brink of Chaos]'. *Corriere Della Sera*, 9 May 2008.
Coward, Martin. 2004. 'Urbicide in Bosnia'. In *Cities, War and Terrorism: Towards an Urban Geopolitics*, edited by Stephen Graham, 154–71. Oxford: Blackwell.
Coward, Martin. 2006. 'Against Anthropocentrism: The Destruction of the Built Environment as a Distinct Form of Political Violence'. *Review of International Studies* 32 (3): 419.
Coward, Martin. 2009. *Urbicide. The Politics of Urban Destruction*. Oxon: Routledge.
Cowen, Deborah, and Neil Smith. 2009. 'After Geopolitics? From the Geopolitical Social to Geoeconomics'. *Antipode* 41 (1): 22–48.
Cunningham, D. E., K. S. Gleditsch, and I. Salehyan. 2013. 'Non-State Actors in Civil Wars: A New Dataset'. *Conflict Management and Peace Science* 30 (5): 516–31.
Davie, Michael. 1983. 'Comment Fait-on La Guerre a Beyrouth?' *Herodote*, 29–30: 17–54.
Davie, Michael. 1992. '"Beyrouth-Est" et "Beyrouth-Ouest". Aux Origines Du Clivage Confessionel de La Ville'. *Les Cahiers d'URBAMA* 8.
Davis, Diane E, and Nora Libertun de Duren. 2011. *Cities & Sovereignty Identity Politics in Urban Spaces*. Bloomington: Indiana University Press.
Demarest, Geoffrey. 1995. 'Geopolitics and Urban Armed Conflict in Latin America'. *Small Wars & Insurgencies* 6 (1): 44–67.
Dīb, Kamāl. 2006. *Warlords and Merchants: The Lebanese Business and Political Establishment*. Reading: Ithaca Press.

Dijink, Gertjan. 2002. *National Identity and Geopolitical Visions: Maps of Pride and Pain*. London: Routledge.
Dodds, Klaus. 2003. 'Licensed to Stereotype: Geopolitics, James Bond and the Spectre of Balkanism'. *Geopolitics* 8 (2): 125–56.
Dodds, Klaus-John, and James Derrick Sidaway. 1994. 'Locating Critical Geopolitics'. *Environment and Planning D: Society and Space* 12 (5): 515–24. https://doi.org/10.1068/d120515.
Douzet, Frédérick. 2001. 'Pour une démarche nouvelle de géopolitique urbaine à partir du cas d'Oakland (Californie)'. *Hérodote* 101 (2): 57. https://doi.org/10.3917/her.101.0057.
Dumper, Mick, and Wendy Pullan. 2010. 'Jerusalem: The Cost of Failure'. Briefing paper MENAP BP 2010/03. London: Chatham House. http://www.chathamhouse.org/sites/default/files/public/Research/Middle%20East/bp0210jerusalem.pdf.
Earle, Lucy. 2016. 'Urban Crises and the New Urban Agenda'. *Environment and Urbanization* 28 (1): 77–86. https://doi.org/10.1177/0956247815620335
Eddé, Carla. 2009. *Beyrouth, Naissance d'une Capitale: 1918–1924*. La Bibliothèque Arabe. Hommes et Sociétés. [Paris]. Arles: Sindbad; Actes sud.
Eisenhower, Dwight. 1958a. 'Special Message to the Congress on the Sending of United States Forces to Lebanon'. Online by Gerhard Peters and John T. Woolley, The American Presidency project. http://www.presidency.ucsb.edu/ws/?pid=11132.
Eisenhower, Dwight. 1958b. 'Statement by the President Following the Landing of United States Marines at Beirut'. *The American Presidency Project*. http:www.presidency.ucsb.edu/ws/?pid=11133.
Elden, Stuart. 2009. *Terror and Territory. The Spatial Extent of Sovereignty*. Minneapolis: University of Minnesota Press.
Eleftériadès, E. 1944. *Les Chemins de Fer En Syrie et Au Liban : Étude Historique, Financière et Économique*. Beyrouth: Impr. Catholique.
El-Khazen, F. 2004. 'Ending Conflict in Wartime Lebanon: Reform, Sovereignty and Power, 1976–88'. *Middle Eastern Studies* 40 (1): 65–84.
El-Khazen, Farid. 2000. *The Breakdown of the State in Lebanon, 1967–1976*. Cambridge, MA: Harvard University Press.
Farah, Caesar E. 2000. *The Politics of Interventionism in Ottoman Lebanon, 1830–1861*. Oxford: Centre for Lebanese Studies [u.a.].
Farish, Matthew. 2004. 'Another Anxious Urbanism: Simulating Defence and Disaster in Cold War America'. In *Cities, War and Terrorism: Towards an Urban Geopolitics*, edited by Stephen Graham, 93–109. Oxford: Blackwell.
Fawaz, Leila Tarazi. 1988. 'Zahle and Daayr Al-Qamar. Two Market Towns of Mount Lebanon during the Civil War of 1860'. In *Lebanon: A History of Conflict and Consensus*, edited by Nadim Shehadi and Dana Haffar Mills, 49. London: I.B. Tauris.
Fawaz, Leila Tarazi. 1994. *An Occasion for War: Civil Conflict in Lebanon and Damascus in 1860*. Berkeley: University of California Press.

Fawaz, Mona. 2009. 'Neoliberal Urbanity and the Right to the City: A View from Beirut's Periphery'. *Development and Change* 40 (5): 827–52.

Fawaz, Mona, and I. Peillen. 2003. '"Urban Slums Reports: The Case of Beirut, Lebanon"', Understanding Slums: Case Studies for the Global Report on Human Settlements'. http://www.ucl.ac.uk/dpuprojects/Global_Report/pdfs/Beirut_bw.pdfhttp://www.ucl.ac.uk/dpupro.

Fawaz, Mona, Mona Harb, and Ahmad Gharbieh. 2012. 'Living Beirut's Security Zones: An Investigation of the Modalities and Practice of Urban Security: Living Beirut's Security Zones'. *City & Society* 24 (2): 173–95. https://doi.org/10.1111/j.1548-744X.2012.01074.x.

Fawaz, Mona, and I Peillen. 2003. '"Urban Slums Reports: The Case of Beirut, Lebanon"', Understanding Slums: Case Studies for the Global Report on Human Settlements'. http://www.ucl.ac.uk/dpuprojects/Global_Report/pdfs/Beirut_bw.pdfhttp://www.ucl.ac.uk/dpupro.

Firro, Kais M. 2003. *Inventing Lebanon: Nationalism and the State under the Mandate*. Library of Middle East History 6. London: I.B. Tauris.

Firro, Kais M. 2006. 'Ethnicizing the Shi'is in Mandatory Lebanon'. *Middle Eastern Studies* 42 (5): 741–59. https://doi.org/10.1080/00263200600827933.

Flint, Colin. 2006. 'Cities, War, and Terrorism: Towards an Urban Geopolitics (Book Review)'. *Annals of the Association of American Geography* 96 (1): 216–18.

Foreign and Commonwealth Office. 1975a. 'Internal Political Situation in Lebanon'. FCO 93/677, 325. National Archives, Kew.

Foreign and Commonwealth Office. 1975b. 'Humanitarian Assistance for Lebanon'. FCO 93/677 Internal political situation in Lebanon. Kew National Archives.

Foreign Mission in Beirut. 1975. 'Internal Political Situation in Lebanon, 18 December'. FCO 93/678 (H.M. diplomatic service, dept Near East and North Africa).

Foucault, Michel. 2000. 'Space, Knowledge and Power (Interview Conducted by Paul Rabinow and Appeared in Skyline, 1982)'. In *Michel Foucault. Essential Works of Foucault 1954–1984 Vol III*, edited by J. Faubion, 349–64. London: Penguin.

Foucault, Michel. 2002. *Archaeology of Knowledge*. Routledge Classics. London; New York: Routledge.

Foucault, Michel. 2003. *Society Must Be Defended: Lectures at the Collège de France, 1975–76*. Nachdr. Lectures at the Collège de France. London: Penguin.

Foucault, Michel. 2007. 'The Meshes of Power'. In *Space, Knowledge and Power. Foucault and Geography*, edited by J. W. Crampton and Stuart Elden, 153–62. Aldershot: Ashgate.

Foucault, Michel, and Colin Gordon. 1980. *Power/Knowledge: Selected Interviews and Other Writings, 1972–1977*. 1st American ed. New York: Pantheon Books.

Fregonese, Sara. 2009a. 'The Urbicide of Beirut? Geopolitics and the Built Environment in the Lebanese Civil War (1975–1976)'. *Political Geography* 28 (5): 309–18.

Fregonese, Sara. 2012a. 'Beyond the 'Weak State': Hybrid Sovereignties in Beirut'. *Environment and Planning D: Society and Space* 30 (4): 655–74.

Fregonese, Sara. 2012b. 'Urban Geopolitics 8 Years on. Hybrid Sovereignties, the Everyday, and Geographies of Peace'. *Geography Compass* 6 (5): 290–303.

Fregonese, Sara. 2012c. 'Between a Refuge and a Battleground: Beirut's Discrepant Cosmopolitanisms'. *Geographical Review* 102 (3): 316–36.

Fregonese, Sara. 2015. 'Everyday Political Geographies'. In *The Wiley-Blackwell Companion to Political Geography*, edited by John Agnew, Virginie Mamadouh, Anna Secor, and Joanne Sharp, 2nd ed, 493–506. London: John Wiley & Sons Ltd.

Frenza, Massimiliano. 2008. 'Libano: L'ultima Cittadella in Partibus Infidelium. Analisi Delle Frammentazioni Politiche Nel Campo Cristiano Maronita e Dei Tentativi Operati Dalla Santa Sede per Riportare l'unità Fra i Cristiani Del Libano'. CeSDIS - Centro Studi per la Difesa e la Sicurezza.

Galtung, Johan. 1969. 'Violence, Peace, and Peace Research'. *Journal of Peace Research* 6 (3): 167–91.

Gendzier, Irene L. 2006. *Notes from the Minefield: United States Intervention in Lebanon and the Middle East, 1945–1958*. New York: Columbia University Press.

Georgelin, Herve. 2003. 'Smyrne à La Fin de l'empire Ottoman : Un Cosmopolitisme Si Voyant'. *Cahiers de La Méditerranée* 67: 125–47.

Giddens Anthony, 1981. *The Nation-state and Violence. Volume Two of a Contemporary Critique of Historical Materialism*. Cambridge: Polity Press.

Giddens, Anthony. 1985. *The Nation State and Violence. Volume Two of a Critique of Historical Materialism*. Oakland: University of California Press.

Giddens, Anthony. 1987. *The Nation-State and Violence. Volume Two of a Critique of Historical Materialism*. A Contemporary Critique of Historical Materialism. Berkeley: Univ. of Calif. Pr.

Glasze, Georg. 2003. 'Segmented Governance Patterns. Fragmented Urbanism: The Development of Guarded Housing Estates in Lebanon'. *The Arab World Geographer* 6 (2): 79–100.

Godlewska, A. 1994. 'Napoleon's Geographers (1797–1815): Imperialists and Soldiers of Modernity'. In *Geography and Empire*, edited by N. Smith, 31–53. Oxford: Blackwell.

Graham, Stephen. 2003. *Lessons in Urbicide*. Vol. 19. *New Left Review* 19 (January).

Graham, Stephen. 2004a. 'Cities as Strategic Sites: Place Annihilation and Urban Geopolitics'. In *Cities, War and Terrorism*, edited by Stephen Graham, 31–53. Oxford: Blackwell.

Graham, Stephen, ed. 2004b. *Cities, War, and Terrorism: Towards an Urban Geopolitics*. Studies in Urban and Social Change. Malden, MA: Blackwell Publishing.

Graham, Stephen. 2004c. 'Epilogue'. In *Cities, War and Terrorism: Towards an Urban Geopolitics*, edited by Stephen Graham, 330–34. Oxford: Blackwell.

Graham, Stephen. 2004d. 'Introduction: Cities, Warfare and States of Emergency'. In *Cities, War and Terrorism: Towards an Urban Geopolitics*, edited by Stephen Graham, 1–26. Oxford: Blackwell.

Graham, Stephen. 2004e. 'Postmortem City: Towards an Urban Geopolitics'. *City* 8 (2): 165–96. https://doi.org/10.1080/1360481042000242148.

Graham, Stephen. 2005. 'Remember Fallujah: Demonising Place, Constructing Atrocity'. *Environment and Planning D: Society and Space* 23 (1): 1–10.

Graham, Stephen. 2006. 'Cities and the "War on Terror"'. *International Journal of Urban and Regional Research* 30 (2): 255–76.

Graham, Stephen. 2008. 'Robowar™ Dreams: US Military Technophilia and Global South Urbanisation'. *City* 12 (1): 25–49.

Graham, Stephen. 2010. *Cities under Siege: The New Military Urbanism*. London; New York: Verso.

Gregory, D. 1995. 'Imaginative Geographies'. *Progress in Human Geography* 19 (4): 447–85.

Gregory, Derek. 2004. *The Colonial Present*. Oxford: Blackwell.

Gregory, Derek. 2006. '"In Another Time Zone, the Bombs Fall Unsafely": Targets, Civilians and Late Modern War'. *Arab World Geographer* 9 (2): 88–111.

Gregory, Derek, and Allan Pred. 2007. Violent Geographies. *Fear, Teror, and Political Violence*. Oxon: Routledge.

Gregory, Derek, Ron Johnston, Geraldine Pratt, Michael Watts, and Sarah Whatmore, eds. 2009. *The Dictionary of Human Geography*. Oxford: Blackwell.

Gregory, Derek. 2010. "War and Peace." *Transactions of the Institute of British Geography* 35: 154–86.

Gurr, Ted Robert. 1994. 'Peoples against States: Ethnopolitical Conflict and the Changing World System: 1994 Presidential Address'. *International Studies Quarterly* 38 (3): 347.

Hakim, Carol. 2013. *The Origins of the Lebanese National Idea, 1840–1920*. Berkeley: University of California Press.

Hanf, Theodor. 2015. *Coexistence in Wartime Lebanon: Decline of a State and Rise of a Nation*. New paperback ed. London: I.B. Tauris.

Haraway, Donna Jeanne. 1991. *Simians, Cyborgs and Women : The Reinvention of Nature*. London: Free Association.

Harb, C. 2008. 'Shock and Awe' in Beirut'. 12 May 2008. http://www.guardian.co.uk/commentisfree/2008/may/12/shockandaweinbeirut.

Harb, Mona, and Lara Deeb. 2009. 'Altre Pratiche Della Resistenza: Turismo Politico e Svaghi Devozionali'. In *Hezbollah: Fatti, Luoghi, Protagonist e Testimonianze*, edited by S. Mervin, 209–26. Milan: Epoche'.

Harik, Judith. 1994. 'The Public and Social Services of the Lebanese Militias'. *Centre for Lebanese Studies*, Papers on Lebanon, no. 14.

Harker, Christopher. 2014. 'The Only Way Is Up? Ordinary Topologies of Ramallah: Ordinary Topologies of Ramallah'. *International Journal of Urban and Regional Research* 38 (1): 318–35.

Harley, Brian. 1998. 'Maps, Knowledge and Power'. In *The Iconography of Landscape*, edited by Dennis Cosgrove and S. Daniels, 277–312. Cambridge: Cambridge University Press.

Harris A, 2008. *Half Full or Half Empty: Assessing Prospects for Peace in Lebanon* (United States Institute of Peace, Washington, DC).

Harris, A. 2015. 'Vertical Urbanisms: Opening Up Geographies of the Three-Dimensional City'. *Progress in Human Geography* 39 (5): 601–20.

Harvey, David. 2008. 'The Right to the City'. *New Left Review* 53.

Haugbølle, Sune. 2010. *War and Memory in Lebanon*. Cambridge: Cambridge University Press.

Henderson, Errol, and J. Singer. 2002. '"New Wars" and Rumors of "New Wars"'. *International Interactions* 28 (2): 165–90. https://doi.org/10.1080/03050620212098.

Hertslet, Sir Edward. 1875. *The Map of Europe by Treaty: Volume II 1828–1863*. London: Butterworths.

Hewitt, Kenneth. 1983. 'Place Annihilation: Area Bombing and the Fate of Urban Places'. *Annals of the Association of American Geographers* 73 (2): 257–84.

Hourani, A. 1976. 'Ideologies of the Mountain and the City'. In *Essays on the Crisis in Lebanon*, edited by Roger Owen, 33–35. London: Ithaca Press.

Hourani, Albert. 1992. *The Lebanese in the World: A Century of Emigration*. London: Centre for Lebanese Studies.

Hourani, Najib. 2010a. 'Transnational Pathways and Politico-Economic Power: Globalisation and the Lebanese Civil War'. *Geopolitics* 15 (2): 290–311.

Hudson, Michael. 1985. *The Precarious Republic: Political Modernisation in Lebanon*. Boulder, CO: Westview Press.

Hulbert, François. 1989. *Essai de Géopolitique Urbaine et Régionale: La Comédie Urbaine de Québec*. Laval, Québec: Éditions du Méridien.

Huxtable, Ada Louise. 1989. *Will They Ever Finish Bruckner Boulevard?* Berkeley: University of California Press.

Hyndman, Jennifer. 2007. '"Feminist Geopolitics Revisited: Body Counts in Iraq*."' *The Professional Geographer* 59 (1): 35–46.

ICRC. 2017. 'War in Cities: What Is at Stake?'. https://www.icrc.org/en/document/war-cities-what-stake-0.

Iossifova, Deljana. 2013. 'Searching for Common Ground: Urban Borderlands in a World of Borders and Boundaries'. *Cities* 34 (October): 1–5.

Isin, Engin. 2002. *Being Political. Genealogies of Citizenship*. Minneapolis: University of Minnesota Press.

İşleyen, Beste. 2018. 'Transit Mobility Governance in Turkey'. *Political Geography* 62 (January): 23–32.

Israel Ministry of Foreign Affairs. 1975. '58 Reply in the Knesset by Defence Minister Peres on Syria's Role in Lebanon'.

Israel Ministry of Foreign Affairs. 1976. 'Statement in the Knesset by Deputy Premier and Foreign Minister Allon on the Situation in Lebanon'.

Jackson, Robert. 1986. 'Negative Sovereignty in Sub-Saharan Africa'. *Review of International Studies* 12 (4): 247–64.

Joseph, Suad. 1975. 'The Politicization of Religious Sects in Borj Hammoud, Lebanon'. PhD dissertation. New York: Columbia University.

Joseph, Suad. 1983. 'Working-Class Women's Networks in a Sectarian State: A Political Paradox'. *American Ethnologist* 10 (1): 1–22.

Kaldor, Mary. 1999. *New and Old Wars : Organized Violence in a Global Era*. Cambridge: Polity Press.

Kassir, Samir. 2003. *Histoire de Beyrouth*. Paris: Fayard.

Kassir, Samir. 2010. *Beirut*. Berkeley: University of California Press.

Kastrinou, A. Maria A. 2016. *Power, Sect and State in Syria: The Politics of Marriage and Identity amongst the Druze*. London; New York: I.B. Tauris.

Kemp, Percy. 1983. 'La Strategie de Bashir Gemayel. Atopie Urbaine et Utopie Territoriale'. *Herodote* 29–30: 55–82.

Khalaf, Samir. 1993. *Beirut Reclaimed: Reflections on Urban Design and the Restoration of Civility*. Beirut: Dar An-Nahar.

Khalaf, Samir. 2002a. *Civil and Uncivil Violence in Lebanon: A History of the Internationalization of Communal Conflict*. New York: Columbia University Press.

Khalaf, Samir. 2002b. *Civil and Uncivil Violence in Lebanon: A History of the Internationalization of Communal Conflict*. New York: Columbia University Press.

Khalaf, Samir, and Philip S. Khoury, eds. 1993. *Recovering Beirut: Urban Design and Post-War Reconstruction*. Social, Economic, and Political Studies of the Middle East [Etudes Sociales, Économiques et Politiques Du Moyen Orient], v. 47. Leiden; New York: Brill.

Khalidy, Soraya. 2003. *Le goût de Beyrouth*. Le petit mercure. Paris: Mercure de France.

Khoury, R., and B. Garfield. 2008. 'Taking Fire: Transscript'. *On the Media* (blog). 2008. www.onthemedia.org/transcripts/2008/05/16/01.

Kobayashi, Audrey. 2017. 'Spatiality'. In *International Encyclopedia of Geography: People, the Earth, Environment and Technology*, edited by Douglas Richardson, Noel Castree, Michael F. Goodchild, Audrey Kobayashi, Weidong Liu, and Richard A. Marston, 1–7. Oxford: John Wiley & Sons, Ltd.

Koopman, Sara. 'Alter-Geopolitics: Other Securities Are Happening'. *Geoforum* 42 (3) (June 2011): 274–84.

Korf, B. 2011. 'Resources, Violence and the Telluric Geographies of Small Wars'. *Progress in Human Geography* 35 (6): 733–56.

Kramer, M. 1993. 'Hizbullah: The Calculus of Jihad'. In *Fundamentalism and the State: Remaking Politics, Economies, and Militance*, edited by M. Marty and R. Appleby, 539–56. Chicago, IL: University of Chicago Press.

Krasner, Stephen D. 1999. *Sovereignty: Organized Hypocrisy*. Princeton, NJ: Princeton University Press.

Krasner, Stephen D. 2001. *Problematic Sovereignty. Contested Rules and Political Possibilities*. New York: Columbia University Press.

Kurgan, Laura. 2017. 'Conflict Urbanism, Aleppo: Mapping Urban Damage'. *Architectural Design* 87 (1): 72–77. https://doi.org/10.1002/ad.2134.

Labaki, Boutros, and Khalil Abou Rjeily. 1993. *Bilan Des Guerres Du Liban, 1975–1990*. Collection 'Comprendre Le Moyen-Orient'. Paris: L'Harmattan.

Lacoste, Yves. 1982. 'D'autres Géopolitiques'. *Hérodote* 25.

Larkin, Craig. 2012. *Memory and Conflict in Lebanon: Remembering and Forgetting the Past*. Exeter Studies in Ethno Politics 3. London; New York: Routledge.

Latour, Bruno. 1993. 'On Technical Mediation: The Messenger Lectures on the Evolution of Civilization'. In *Institute of Economic Research, Working Papers Series*. Ithaca, NY: Cornell University, Institute of Economic Research.

Launchbury, Claire. 2014. 'The Impossible Archive of Beirut'. *Francosphères* 3 (1): 99–113. https://doi.org/10.3828/franc.2014.7.

Le Monde. 2008. 'Beyrouth Replonge Dans La Guerre Des Rues,' 9 May 2008.

League of Nations. 1945. 'The Mandates System. Origin, Principles, Application'. LoN/1945.VI.A.1. https://unispal.un.org/DPA/DPR/unispal.nsf/0/C61B138F4DBB08A0052565D00058EE1B.

Leontidou, Lila. 2001. 'Attack on the Landscape of Power. An Anti-War Elegy to New York Inspired by Whitman's Verses'. *City* 5 (3): 406–10.

LIFE Magazine. 1958. 'Cold War Moves to a Showdown'. *LIFE Magazine*, 28 July 1958.

Lindqvist, Sven. 2002. *A History of Bombing*. London: Granta.

Lindqvist, Sven. 2003. *A history of bombing*. New York: The New Press.

Lisle, Debbie. 2016. *Holidays in the Danger Zone: Entanglements of War and Tourism*. Critical War Studies. Minneapolis: University of Minnesota Press.

Little, Douglas. 1996. 'His Finest Hour? Eisenhower, Lebanon, and the 1958 Middle East Crisis'. *Diplomatic History* 20 (1): 27–54.

L'Orient le Jour. 1975a. 'Brahim Koleilat: Notre Riposte a Empeche Une Contre-Saint Barthelemy', Décembre 1975.

L'Orient le Jour. 1975b. 'De Kantari à Sodeco Beyrouth s'embrase', 9 December 1975.

L'Orient le Jour. 1976. 'Tous Les Fronts s'embrasent', 25 March 1976.

Lunstrum, Elizabeth. 2013. 'Articulated Sovereignty: Extending Mozambican State Power through the Great Limpopo Transfrontier Park'. *Political Geography* 36 (September): 1–11.

Maasri, Zeina. 2008. *Off the Wall: Political Posters of the Lebanese Civil War*. London: I.B. Tauris.

MacDonald, Fraser, Rachel Hughes, and Klaus Dodds, eds. 2010. *Observant States: Geopolitics and Visual Culture*. International Library of Human Geography 16. London; New York: I.B. Tauris.

Majed, Z. 1996. 'Que Reste T-Il Du Hezbollah'. *L'Orient-Express*, November 1996.

Makdisi, Saree. 1997. 'Laying Claim to Beirut: Urban Narrative and Spatial Identity in the Age of Solidere'. *Critical Inquiry* 23 (3): 660–705.

Makdisi, Ussama. 2000. *The Culture of Sectarianism Community, History, and Violence in Nineteenth-Century Ottoman Lebanon*. Berkeley, CA: University of California Press.

Marcuse, Peter. 2004. 'The "War on Terrorism" and Life in Cities after September 11, 2001.' In *Cities, War and Terrorism: Towards an Urban Geopolitics*, edited by Stephen Graham, 263–75. Oxford: Blackwell.
Martinez, B., and F. Volpicella. 2008. 'Walking the Tight Wire: Conversations on the May 2008 Lebanese Crisis'. Transnational Institute. http://www.tni.org/article/walking-tight-wire.
Mavroudi, Elizabeth. 2010. 'Imagining a Shared State in Palestine-Israel'. *Antipode* 42 (1): 152–78.
McConnell, Fiona. 2009. 'De Facto, Displaced, Tacit: The Sovereign Articulations of the Tibetan Government-in-Exile.' *Political Geography* 28 (6): 343–52.
McConnell, Fiona, Nick Megoran, and Philippa Williams, eds. 2014. *Geographies of Peace*. London: I.B. Tauris.
McElroy, D. 2008. 'Hizbollah "Ready for War" in Lebanon'. *The Telegraph*, 8 May 2008. http://www.telegraph.co.uk/news/worldnews/middleeast/lebanon/1938944/Hizbollah-ready-for-warin-Lebanon.html.
McLeod, Hugh. 2008. 'Lebanese Declaration Threatens Civil War'. *The Observer*, 11 May 2008.
Megoran, Nick. 2010. 'Towards a Geography of Peace: Pacific Geopolitics and Evangelical Christian Crusade Apologies: Towards a Geography of Peace'. *Transactions of the Institute of British Geographers* 35 (3): 382–98.
Megoran, Nick. 2011. 'War and Peace? An Agenda for Peace Research and Practice in Geography'. *Political Geography* 30 (4): 178–89. https://doi.org/10.1016/j.polgeo.2010.12.003.
Mermier, Franck, and Christophe Varin, eds. 2010. *Mémoires de Guerres Au Liban: 1975–1990*. 1re édn. La Bibliothèque Arabe. Hommes et Sociétés. Arles : Paris: Actes sud; Sindbad.
Merrill, Dennis. 2009. *Negotiating Paradise. U.S Tourism and Empire in Twentieth-Century Latin America*. Chapel Hill: University of North Carolina Press.
Minca, Claudio, and Luiza Bialasiewicz. 2004. *Spazio e politica: riflessioni di geografia critica*. Padova: CEDAM.
Ministère des Affaires étrangères. 1976a. 'Déclaration de M. Sauvagnargues Lors Du Débat de Politique Étrangère Au Sénat (Extraits)'.
Ministère des Affaires étrangères. 1976b. 'Interview de M. Sauvagnargues Pour TF1 Sur Le Liban'. Paris.
Ministère des Affaires étrangères. 1976c. 'Discours de M. de Guiringaud Devant La Xxxie Assemblee Generale Des Nations Unies'.
Mitchell, Timothy. 1988. *Colonising Egypt*. Berkeley: University of California Press.
Moisio, Sami. 2015. 'Geopolitics/Critical Geopolitics'. In *The Wiley Blackwell Companion to Political Geography*, edited by John Agnew, Virginie Mamadouh, Anna Secor, and Joanne Sharp, 220–34. Oxon: Blackwell.
Monroe, Kristin V. 2016. *The Insecure City: Space, Power, and Mobility in Beirut*. New Brunswick, NJ: Rutgers University Press.

Morrissey, Mike, and Frank Gaffikin. 2006. 'Planning for Peace in Contested Space'. *International Journal of Urban and Regional Research* 30 (4): 873–93. https://doi.org/10.1111/j.1468-2427.2006.00696.x.

Mörtenböck, Peter, and Helge Mooshammer, eds. 2008. *Networked Cultures: Parallel Architectures and the Politics of Space*. Rotterdam; New York: NAi Publishers; D.A.P./Distributed Art Publishers (distributor).

Mostar Architects Association. 1993. 'Mostar '92 – Urbicide'. *Spazio e Società/ Space and Society* 16 (62): 8–25.

Mountz, A. 2013. 'Political Geography I: Reconfiguring Geographies of Sovereignty'. *Progress in Human Geography*, March. https://doi.org/10.1177/0309132513479076.

Möystad, Ole. 1998. 'Morphogenesis of the Beirut Green-Line : Theoretical Approaches between Architecture and Geography (note)'. *Cahiers de géographie du Québec* 42 (117): 421. https://doi.org/10.7202/022766ar.

Mufti, Malik. 1996. *Sovereign Creations: Pan-Arabism and Political Order in Syria and Iraq*. Ithaca, NY: Cornell University Press.

Muñoz, J. 2008. 'Hezbolá Reaviva La Violencia En Líbano [Hezbollah Revives Violence in Lebanon]'. *El Pais*, 9 May 2008. http://www.elpais.com/articulo/internacional/Hezbola/reaviva/violencia/Libano/elpepiint/20080509elpepiint_1/Tes.

Naeff, Judith. 2018. *Precarious Imaginaries of Beirut – A City's Suspended Now*. Palgrave Macmillan. www.palgrave.com/la/book/9783319659329.

Nagel, C. 2002. 'Reconstructing Space, Re-Creating Memory: Sectarian Politics and Urban Development in Post-War Beirut'. *Political Geography* 21 (5): 717–25. https://doi.org/10.1016/S0962-6298(02)00017-3.

Naharnet. 2008. 'Iran Points Fi Nger at US and Israel, Assad Says Lebanon Unrest 'Internal Matter', 9 May 2008.

Nasr, Salim. 1990. 'Lebanon's War Is the End in Sight?' *Merip Report* 20 (162). http://www.merip.org/mer/mer162/primer-lebanons-15-year-war-1975-1990.

Nordstrom, Carolyn, and Antonius C. G. M. Robben, eds. 1995. *Fieldwork under Fire: Contemporary Studies of Violence and Survival*. Berkeley: University of California Press.

Norton, Augustus R. 2014. *Hezbollah: A Short History*. New paperback ed. Princeton Studies in Muslim Politics. Princeton, NJ: Princeton University Press.

Nucho, Joanne. 2016. *Everyday Sectarianism in Urban Lebanon: Infrastructures, Public Services, and Power*. Princeton Studies in Culture and Technology. Princeton, NJ: Princeton University Press.

Ó Tuathail, Gearoid. 1992. 'Foreign Policy and the Hyperreal: The Reagan Administration and the Scripting of "South Africa"'. In *Writing Worlds: Discourse, Text and Metaphor in the Representation of Landscape*, edited by Trevor J. Barnes and James S. Duncan, 155–75. Oxon: Routledge.

Ó Tuathail, Gearóid. 1996a. 'An Anti-Geopolitical Eye: Maggie O'Kane in Bosnia, 1992–93'. *Gender, Place and Culture* 3 (2): 171–85.

Ó Tuathail, Gearóid. 1996b. *Critical Geopolitics*. London: Routledge.

Ó Tuathail, Gearóid. 1996. *Critical Geopolitics: The Politics of Writing Global Space*. 1st ed. London: Routledge.
Ó Tuathail, Gearóid. 2002. 'Theorizing Practical Geopolitical Reasoning: The Case of the United States' Response to the War in Bosnia'. *Political Geography* 21 (5): 601–28. https://doi.org/10.1016/S0962-6298(02)00009-4.
Paul VI. 1975a. 'Discours Du Pape Paul VI Au Nouvel Ambassadeur de La République Libanais Près Le Saint Siège'.
Paul VI. 1975b. 'Message Du Pape Paul VI à s.e. Monsieur Soleiman Frangié, Président de La République Libanaise'.
Press, D. 1998. 'Urban Warfare: Options, Problems, and the Future'. In *Proceedings of the MIT Security Studies Conference Program*. Officer's Club. Hanscom Air Force Base. Bedford, Massachussets.
Pullan, Wendy. 2006. 'Locating the Civic in the Frontier: Damascus Gate'. In *Did Someone Say Participate?: An Atlas of Spatial Practice*, edited by Markus Miessen and Shumon Basar. Cambridge, MA: MIT Press.
Pullan, Wendy. 2011. 'Frontier Urbanism: The Periphery at the Centre of Contested Cities'. *The Journal of Architecture* 16 (1): 15–35. https://doi.org/10.1080/13602365.2011.546999.
Pullan, Wendy, Philipp Misselwitz, Rami Nasrallah, and Haim Yacobi. 2007. 'Jerusalem's Road 1'. *City* 11 (2): 176–98. https://doi.org/10.1080/13604810701395993.
Qassem, Naim. 2012. *Hizbullah: The Story from Within*. London: Saqi. http://public.eblib.com/choice/publicfullrecord.aspx?p=978765.
Ramadan, A. 2009. 'Destroying Nahr El-Bared: Sovereignty and Urbicide in the Space of Exception'. *Political Geography* 28 (3): 153–63.
Ramadan, Adam, and Sara Fregonese. 2017. 'Hybrid Sovereignty and the State of Exception in the Palestinian Refugee Camps in Lebanon'. *Annals of the American Association of Geographers* 107 (4): 949–63.
Ramet, Sabrina Petra. 1996. 'Nationalism and the "Idiocy" of the Countryside: The Case of Serbia'. *Ethnic and Racial Studies* 19 (1): 70–87.
Reagan, Ronald. 1983. 'President Reagan's Televised Address to the United States of America, Speaking of Both the Terrorist Bombing in Lebanon and the Grenada Invasion'. The Beirut memorial online. www.beirut-memorial.org/history/reagan.html.
Reid-Henry, Simon, and Ole Jacob Sending. 2014. 'The "Humanitarianization" of Urban Violence'. *Environment and Urbanization* 26 (2): 427–42. https://doi.org/10.1177/0956247814544616.
Rodogno, Davide. 2012. *Against Massacre: Humanitarian Interventions in the Ottoman Empire, 1815–1914: The Emergence of a European Concept and International Practice*. Human Rights and Crimes against Humanity. Princeton, NJ: Princeton University Press.
Rogan, Eugene L. 2016. *The Fall of the Ottomans: The Great War in the Middle East, 1914–1920*. Penguin History. London: Penguin Books.

Rokem, J., and Boano, eds. 2017. *Urban Geopolitics Rethinking Planning in Contested Cities*. Taylor & Francis.
Rose, Gillian. 2001. *Visual Methodologies: An Introduction to the Interpretation of Visual Materials*. 1st ed. London: Sage.
Rowayheb, Marwan G., and Makram Ouaiss. 2015. 'The Committee of the Parents of the Missing and Disappeared: 30 Years of Struggle and Protest'. *Middle Eastern Studies* 51 (6): 1010–26.
Rowayheb, Marwan George. 2006. 'Lebanese Militias: A New Perspective'. *Middle Eastern Studies* 42 (2): 303–18.
Roy, A. 2009. 'Civic Governmentality: The Politics of Inclusion in Beirut and Mumbai'. *Antipode* 41 (1): 159–79.
Ruppert, Helmut, and Éric Verdeil. 1999. 'Beyrouth, une ville d'orient marquée par l'occident'. *Mitteilungen der Fränkischen Geographischen Gesellschaft; Bd. 15/16*, Les cahiers du CERMOC.
Saadé, Joseph, Frédéric Brunnquell, and Frédéric Couderc. 1989. *Victime et bourreau: une vie*. Paris: Calmann-Lévy.
Safier, Michael. 2001. 'Confronting 'Urbicide': Crimes against Humanity, Civility and Diversity, and the Case for a Civic Cosmopolitan Response to the Attack on New York'. *City* 5 (3): 416–29.
Said, Edward W. 1994. *Culture and Imperialism*. London: Vintage.
Said, Edward W. 1995. *Orientalism*. Harmondsworth: Penguin.
Salam, N. 2001. *La Condition Libanaise: Communautés, Citoyen, Etat*. Beirut: Dar an-Nahar.
Salam, Nawaf. 1979. 'L'insurrection de 1958 Au Liban'. Thesis 3e cycle Hist, Paris 4.
Salem, Paul. 1992. 'Superpowers and Small States : American-Lebanese Relations in Perspective'. *Cahiers de la Méditerranée* 44 (1): 135–64.
Saliba, Robert, and Michel Assaf. 1998. *Beirut 1920–1940: Domestic Architecture between Tradition and Modernity*. Beirut: Order of Engineers and Architects.
Salibi, Kamal S. 1976. *The Modern History of Lebanon*. Westport, CT: Greenwood Press.
Salibi, Kamal S. 2003. *A House of Many Mansions: The History of Lebanon Reconsidered*. London: I.B. Tauris.
Sarkis, J. (1993) *Histoire de la guerre du Liban 1975–1990*. Paris: Presses Universitaires de France.
Sassen, Saskia. 2005. 'The Repositioning of Citizenship and Alienage: Emergent Subjects and Spaces for Politics'. *Globalizations* 2 (1): 79–94.
Sassen, Saskia. 2010. 'When the City Itself Becomes a Technology of War'. *Theory, Culture & Society* 27 (6): 33–50. https://doi.org/10.1177/0263276410380938.
Sassen, Saskia. 2018. 'Welcome to a New Kind of War: The Rise of Endless Urban Conflict'. 30 January 2018. https://www.theguardian.com/cities/2018/jan/30/new-war-rise-endless-urban-conflict-saskia-sassen.

Sauvagnargues. 1976. 'Lettre Sur Le Liban Adressée Par M. Sauvagnargues à M. Aldheim, Secretair Général Des Nations Unies, En Response a Celle de m. Bouteflika, Ministre Algérien Des Affaires Étrangères'. Mai 1976.
Sawalha, Aseel. 1998. 'The Reconstruction of Beirut: Local Responses to Globalization'. *City & Society* 10 (1): 133–47.
Sawalha, Aseel. 2000. 'Post-War Beirut: Place Attachment and Interest Groups in Ayn Al-Mreisi'. *The Arab World Geographer* 3 (4): 289–302.
Sawalha, Aseel. 2010. *Reconstructing Beirut: Memory and Space in a Postwar Arab City*. 1st ed. Jamal and Rania Daniel Series in Contemporary History, Politics, Culture, and Religion of the Levant. Austin: University of Texas Press.
Sayigh, Anis. 1955. *Lubnan Al-ṭā'ifi (Sectarian Lebanon)*. Beirut: Dar al-Sira' al-Fikri.
Sayigh, Rosemary. 1977. 'Sources of Palestinian Nationalism: A Study of a Palestinian Camp in Lebanon'. *Journal of Palestine Studies* 6 (4): 17–40.
Sayigh, Rosemary. 1994. *Too Many Enemies*. London: Zed.
Sayigh, Yezid. 1997. 'Armed Struggle and State Formation'. *Journal of Palestine Studies* 26 (4): 17–32.
Schlöndorff, Volker. 1981. *Die Fälschung*. Kino International.
Sebald, W. G., and Anthea Bell. 2004. *On the Natural History of Destruction*. Modern Library pbk. ed. New York: Modern Library.
Shapiro, Michael J. 1997. *Violent Cartographies: Mapping Cultures of War*. Minneapolis: University of Minnesota Press.
Sharp, Joanne. 1998. 'Reel Geographies of the New World Order. Patriotism, Masculinity, and Geopolitics in Post-Cold War American Movies'. In *Rethinking Geopolitics*, edited by Simon Dalby and Gearoid Ó Tuathail, 152–69. London: Routledge.
Shaw, Martin. 2004. 'New Wars of the City: Relationships of "Urbicide" and "Genocide"'. In *Cities, War, and Terrorism*, edited by Stephen Graham, 141–53. Malden, MA: Blackwell Publishing.
Shehadi, Nadim. 2008. 'Lebanon: Futile Victory'. *The World Today* 64 (June).
Shehadi, Nadim, and Dana Haffar Mills, eds. 1988. *Lebanon: A History of Conflict and Consensus*. London: Centre for Lebanese Studies in association with I.B. Tauris.
Shukrallah Haidar, Antoine. 1975. 'Beirut 2000. The Ugly Face of Lebanon'. *Al Hawadess*, 28 February 1975.
Sidaway, James. 2003. 'Sovereign Excesses? Portraying Postcolonial Sovereigntyscapes'. *Political Geography* 22: 157–78.
Sidaway, James D., Till F. Paasche, Chih Yuan Woon, and Piseth Keo. 2014. 'Transecting Security and Space in Phnom Penh'. *Environment and Planning A* 46 (5): 1181–1202. https://doi.org/10.1068/a46167.
Simmons, C. 2001. *Urbicide and the Myth of Sarajevo*. Vol. 68.
Smith, Neal. 2006. 'Cities, War, and Terrorism: Towards an Urban Geopolitics (Book Review)'. *International Journal of Urban and Regional Studies* 30 (2): 469–70.

Soja, Edward W. 2000. *Postmetropolis: Critical Studies of Cities and Regions*. Malden, MA: Blackwell Pub.

Sorkin, Michael, and Stephen Graham. 2004. 'Urban Warfare. A Tour of the Battlefield'. In *Cities, War and Terrorism: Towards an Urban Geopolitics*, edited by Stephen Graham, 251–62. Oxford: Blackwell.

Stel N M, 2009. '"Forcing the Lebanese back to dialogue': Hezbollah's role in the May 2008 Beirut clashes analysed from a state-building perspective" MA thesis, Faculty of Humanities, Universiteit Utrecht.

Stewart, Desmond. 1958. *Turmoil in Beirut: A Personal Account*. London: Wingate.

Sundberg, R., K. Eck, and J. Kreutz. 2012. 'Introducing the UCDP Non-State Conflict Dataset'. *Journal of Peace Research* 49 (2): 351–62.

Thompson, C. W. 2012. *French Romantic Travel Writing: Chateaubriand to Nerval*. Oxford; New York: Oxford University Press.

Time Magazine. 1975. 'Shards from a Shattered Mosaic'. 10 November 1975. http://content.time.com/time/magazine/article/0,9171,913676,00.html.

Tuathail, Gearóid Ó, and John Agnew. 1992. 'Geopolitics and Discourse'. *Political Geography* 11 (2): 190–204.

Turki, Fawaz. 1988. *Soul in Exile: Lives of a Palestinian Revolutionary*. New York: Monthly Review Press.

Tyner, James A, Samuel Henkin, Savina Sirik, and Sokvisal Kimsroy. 2014. 'Phnom Penh during the Cambodian Genocide: A Case of Selective Urbicide'. *Environment and Planning A* 46 (8): 1873–91.

UN Security Council. 1958. 'UN Security Council Meeting, 15 July 1958: Opening Statement by Mr Lodge'. FO 371/134130/1015/498. The National Archives, Kew.

Vignal, Leïla. 2014. 'Destruction-in-Progress: Revolution, Repression and War Planning in Syria (2011 Onwards)'. *Built Environment* 40 (3): 326–41. https://doi.org/10.2148/benv.40.3.326.

Watson, Iain. 2013. '(Re) Constructing a World City: Urbicide in Global Korea'. *Globalizations* 10 (2): 309–25.

Weizman, Eyal. 2003. 'The Evil Architects Do. Crimes of Urbicide and the Built Environment'. In *Content*, edited by Rem Koolhaas.

Weizman, Eyal. 2004. 'Strategic Points, Flexible Lines, Tense Surfaces, and Political Volumes: Ariel Sharon and the Geometry of Occupation'. In *Cities, War, and Terrorism : Towards an Urban Geopolitics*, edited by Stephen Graham, 172–191. Malden, MA: Blackwell.

Weizman, Eyal. 2007. *Hollow Land: Israel's Architecture of Occupation*. London: Verso.

Wharton, Annabel. 2004. *Building the Cold War: Hilton International Hotels and Modern Architecture*. Chicago, IL: Chicago University Press.

Whatmore, Sarah. 2002. *Hybrid Geographies: Natures, Cultures, Spaces*. London; Thousand Oaks, CA: Sage.

White, Hayden V. 1987. *The Content of the Form: Narrative Discourse and Historical Representation*. Baltimore: Johns Hopkins University Press.

White House. 1976a. 'Cabinet Meeting, President Ford and the Cabinet'. Box 19, National Security Adviser. Memoranda of Conversations. Gerald R. Ford Presidential Library.

White House. 1976b. 'Cabinet Meeting, President Ford and the Cabinet. Memorandum of Conversation'. Box 19, National Security Adviser. Memoranda of Conversations. Gerald R. Ford Presidential Library. https://www.fordlibrarymuseum.gov/library/document/0314/1553476.pdf.

White House. 1976c. 'Meeting of the National Security Council – Lebanon'. Box 2, folder: 'NSC Meeting, 4/7/1976' of the National Security Adviser's NSC Meeting File. Gerald R. Ford Presidential Library.

Williams, Philippa. 2015. *Everyday Peace? Politics, Citizenship and Muslim Lives in India*. RGS-IBG Book Series. Chichester; Malden, MA: Wiley Blackwell.

Wright, Paul. Valedictory dispatch. 1975. 'Valedictory Despatch on Lebanon'. 28 April 1975. FCO 93/679. The National Archives.

Yacobi, Haim, and Wendy Pullan. 2014. 'The Geopolitics of Neighbourhood: Jerusalem's Colonial Space Revisited'. *Geopolitics* 19 (3): 514–39.

Yacoub, Gebran. 2010. *Histoire de l'architecture au Liban, 1875–2010* [A History of Architecture in Lebanon, 1875–2010] [خيرات ﺔﺳدنهلا ﺔيرامعم يف لبنان,, *1875–2010*]. Beyrouth: Alphamedia.

Yassin, Nasser. 2010. 'Violent Urbanization and Homogenization of Space and Place: Reconstructing the Story of Sectarian Violence in Beirut (No. 2010, 18)'. Working paper 18, World Institute for Development Economics Research.

Yiftachel, Oren. 1998. 'Planning and Social Control: Exploring the Dark Side'. *Journal of Planning Literature* 12 (4): 395–406.

Yiftachel, Oren, and Haim Yacobi. 2003. 'Urban Ethnocracy: Ethnicization and the Production of Space in an Israeli "Mixed City"'. *Environment and Planning D: Society and Space* 21 (6): 673–93.

Zamir, Meir. 2005. 'An Intimate Alliance: The Joint Struggle of General Edward Spears and Riad Al-Sulh to Oust France from Lebanon, 1942–1944'. *Middle Eastern Studies* 41 (6): 811–32.

Zisser, E. 1997. 'Hizballah at the Crossroads'. In *Religious Radicalism in the Greater Middle East*, edited by B. Weitzman and E. Inbar, 90–110. London: Frank Cass.

Zisser, Eyal. 2003. 'The Mediterranean Idea in Syria and Lebanon: Between Territorial Nationalism and Pan-Arabism'. *Mediterranean Historical Review* 18 (1): 76–90.

# Afterword

## Klaus Dodds

It is a great pleasure to write an afterword to this erudite interrogation of urban geopolitics, using Beirut as a detailed case study. To describe it as a 'case study' is not really to do justice to the care and commitment the author brings to her scholarship. As we learn, Sara Fregonese's (henceforth Sara) connections to Beirut started in her undergraduate years as part of a programme of Arabic language training and have since then been a near-constant companion in her scholarly research into urban geopolitics, taking in along the way the material infrastructure of the city as well as the everyday and affectual qualities of this remarkable port city on the edge of the Eastern Mediterranean.

Before acting as Sara's mentor in 2008 when she held a British Academy Post-Doctoral Fellowship, I had visited Beirut once in the summer of 2003. It was an emotive journey. I was travelling with my young son and my wife to meet her Armenian family who ended up in Beirut after a period of time in the Syrian city of Aleppo. After being met at the airport by my wife's aunt, we travelled to Beit Mery, a small town some 15 kilometres or so from central Beirut. My wife has spent her formative years in Lebanon before the civil war of the 1970s persuaded her family to emigrate to the United States. She had not seen Beit Mery since the 1970s. As we arrived at the family apartment, it was apparent that façade of the building still bore witness to the civil war years. Bullet holes punctuated the brick work.

Once inside the apartment, my wife was reunited with her grandfather, Manuel Vahe Kizirian. He was approximately 100 years old and a survivor of the Armenian genocide. His first great grandson was presented to him, and the photograph of the two of them together is one of my favourites. A little later on in our visit, Manuel took me to his study and we pulled out an atlas. He began to trace with his fingers his life journey from Ottoman Turkey, to Syria (where my wife's father was born) and to finally Lebanon. He trained as a pharmacist and married a fellow Armenian who hailed from another survivor family. As he spoke about his memories of Armenia and then later Beirut, he slipped in and out of multiple languages – speaking to me in English, French, Armenian and Arabic. He admitted he also remembered some of his Ottoman Turkish as well. And as he expanded upon his memories of a changing and changeable Beirut, he took me to his balcony, and beckoned me to enjoy the view of the city and Mediterranean. The hot summer haze obscured the view but it was stunning nonetheless.

The testimony of my wife's grandfather (he died shortly after our visit in 2003) informed my reading of this subtle and detailed exploration of urban politics. The apartment building in Beit Mery was not at the epicentre of the civil wars of the 1970s, but it certainly was implicated in the 1860 Mount Lebanon civil war. As Sara's book suggests, the historical geographies of Beirut and Mount Lebanon need to be carefully excavated in order to make sense of a more contemporary rendering of urban geopolitics. As the book shows, the ways of seeing Lebanon have been varied ranging from descriptors such as 'Paris or Switzerland of the Middle East' to quagmire and disaster zone. For those living in the 1950s and 1960s, especially European and American visitors, Beirut was a place where one could sunbathe in the morning and ski in the afternoon. It looked and felt like a hedonistic paradise. When travelling to Hong Kong, James Bond's writer Ian Fleming wrote in *Thrilling Cities* that Beirut was the 'great smuggling junction of the world' and spoke of how when the aircraft door opened at the airport the 'first sticky fingers of the East reached in … in the pretentious, empty airport' (Fleming 1959, 4–5).

But it was also a time of Cold War tension. While Fleming was alluding to bazaars and a Lebanese penchant for illicit trade, President Eisenhower was more concerned with Soviet intentions in the Eastern Mediterranean. The then president of Lebanon, Camille Chamoun called for the United States' help in the wake of fears that local expressions of strife might be the start of something more sinister. For three months, US Marines in the summer of 1958 occupied the airport and city, and the operation was designed to discourage a fusion of Arab pan-nationalism and Soviet opportunism. Paradoxically, the residents of the city might have thought the arrival of the US Marines was more routine because public communication about the operation had been highly restricted. In the end, the United States lost more men through accidental drowning then they did courtesy of hostile fire. But the legacy of the operation was significant; America positioned itself as a welcomed supporter of pro-Western governments in the Middle East. Others, on the ground in Beirut, were less sanguine about the intervention, and did care for the idea that Lebanon and the Lebanese were bedevilled with collective and communal self-destruction.

So one of the challenges that *War and the City* takes on is the contested geographies of the city, the disputed memories of civil war and external intervention, and the manner in which the city of Beirut was and is shaped by the everyday experiences of conflict and post-conflict. Lest it be forgotten this extraordinary city has also attracted a medley of Western and non-Western characters some of whom ended up being high-profile detainees and kidnapping victims during the civil war years, as well as others such as Osama bin Laden who was alleged to have been inspired by the burning high-rise buildings of the city. In October 2004, Bin Laden shared this extract of a longer speech with the Al-Jazeera television network:

> God knows it did not cross our minds to attack the Towers, but after the situation became unbearable – and we witnessed the injustice and tyranny of the American-Israeli alliance against our people in Palestine and Lebanon – I thought about it. And the events that affected me directly were those of 1982 and the events that followed – when America allowed the Israelis to invade Lebanon, helped by the US Sixth Fleet. As I watched the destroyed towers in Lebanon, it occurred to me to punish the unjust the same way: to destroy towers in America so it could taste some of what we were tasting and to stop killing our children and women. (cited in the *Guardian* 2004)

The residents of Beirut have had to endure a great many things, emotionally, ideologically, materially, and affectively.

If Beirut and Lebanon are a junction, to use Ian Fleming's term, then perhaps our interest is better directed towards how it informed an interdisciplinary set of conversations about urban geopolitics, including its excesses such as destruction and death. In drawing attention to Beirut, Sara shows us how any urban geopolitics worth its salt needs to do three things. First, it must be attentive to the lived and everyday experiences of those who occupy the urban. Second, it must be attentive to the micro-geographies of the urban including its infrastructures such as roads, airports and hotels. The fabric of the city in other words. Finally, it needs to remain sensitive to the embodied experiences and memories of those who were caught up in the civil wars and post-conflict period. Sara's interviewees provide a striking and necessary counter-foil to those who wish to focus on either the destruction of the urban fabric and/or the politics of external intervention and the disruptive politics of regional and global superpowers.

My favourite part of *War and the City* is Chapters 5 and 6 because here we read about the fracturing and fissuring of Beirut due to geopolitical machinations, uneven development and spatial inequalities. The subject matter is grim but the accounting of it matters. Sara shows that the social-spatial complexities of the city do not just make themselves felt during the civil wars of the 1970s. Rather the things that make conflict possible during this period are entirely understandable; Lebanon found itself accommodating not only scores of Palestinians exited from the late 1940s onwards but also adapting to Cold War and regional geopolitical pressures including a proximity to the northern border of Israel and the disputed Golan Heights/

Al-Jawlān. It was easy to forget that Lebanon was home to the Palestine Liberation Organization (PLO) between the late 1960s and early 1980s, and as a consequence of its growing presence in southern Lebanon it attracted an invading Israeli force in 1982. The PLO relocated to Tunisia but the people of Lebanon did not have the luxury of relocation.

But the hostility Israel and the PLO felt towards one another only takes us so far. As the civil war took hold of the country, Lebanon attracts the same kind of place-based characterization that we were later to hear and read for Bosnia in the 1990s. Cosmopolitan cities and spaces become, all to readily, reduced to unrelenting complexity and confusion. Intervene at your peril. The city's built infrastructure and its popular culture become terrains of contention and contentious terrains. Assertions are made about the state of Beirut and things and objects from hotel buildings to political posters attract contentious claims about the past, present and future. Remarkably, for all the disruption and violence amongst rival militias, the city and the state demonstrate a remarkable degree of improvisation. Urban geopolitics is shown to be anything but fixed.

In the last chapter, *War and the City* returns us to the conceptual mission. The intersection of sovereignty, territory and governance have been shown to defy simple characterization. Sara introduces us to the notion of hybrid sovereignties, which seems to me entirely in keeping with the spirit of the book. A concern, in other words, for how the expressions of power and conflict can and do get assembled and reassembled over time and space. Sovereignty in a city like Beirut becomes hybrid because of an assemblage of state and non-state actors. Where the authority of the Lebanese state begins and ends is a moot point. It calls into question earlier framings of Lebanon as lacking sovereignty. Instead the reader is reminded that even in the darkest days of the 1970s there were localized arrangements in place to ensure that life could still function; refuge was collected, infrastructure repaired and maintained and traffic managed. But these hybrid sovereignties can also be disconcerting for others, especially countries such as Israel and Turkey. What emerges, overall, is a more nuanced sense of Beirut and Lebanon; moving away from the idea of a 'failed state' and urban chaos to something more complex involving spatial and political reorganization of national and city life. As Sara's respondents make clear, life could be tough but in their everyday lives we find clues as to how resilience makes itself manifest.

Being resilience is tough in a city divided by civil war and enduring Green Line. Urban planning is a bricolage. State and non-state actors and structures blur and blend with one another. Religious, political and economic interest groups put into action their own agendas, for either making money or exerting control. While the civil wars remind us that the firing of sniper rifles and the dropping of bombs matter, it is equally important to trace through the impact and legacy of more mundane matters such as infrastructure maintenance and protection and planning projects. Some of the more recent disputes and controversies have often revolved around banal objects such as containers and road maintenance.

This is a book to be judged in its own terms. It is an intervention into urban geopolitics and the manner in which the terrain of the urban exposes geopolitical manifestations and ramifications. It does not offer a definitive history of Beirut and Lebanon; rather it shows us how it is possible to do 'urban geopolitics', which is respectful of and sensitive to the histories and geographies of a city striated by civil war and Cold War geopolitics. If Osama bin Laden could take grim inspiration from the civil war affecting Lebanon in 1982, the United States was radicalized by the disappearance and death of the head of CIA station in Beirut, William Buckley. So much of what followed subsequently across the cities of the Middle East and Central Asia has its origins in Beirut in the early 1980s.

Finally, what is abundantly clear from reading *War and the City* is that Sara cares about the city and its inhabitants. The city is not a test-bed for urban geopolitical theorizing but a space of and for curiosity and dignity. It is a book to be read alongside a graphic novel such as *I Remember Beirut*, which provides a vivid depiction of what it was like to be a child at the time of the civil wars (Abirached 2014). The writing and visual registers are different but are trying to make sense of a remarkable city and country.

## References

Abirached, Z. 2014. *I Remember Beirut*. Minneapolis: Graphics Universe.
Bin Laden, O. 2004. 'God Knows It Did Not Cross Our Minds to Attack the Towers'. *The Guardian* 30 October and URL available at: https://www.theguardian.com/world/2004/oct/30/alqaida.september11.
Fleming, I. 1959. *Thrilling Cities*. London: Pan Books.

# Index

Agnew, John 88, 108, 129
Al-Mourabitun militia 81, 110, 116, 118–19, 156 n.3
  propaganda poster by 119–22
Al Mustaqbal newspaper building, attack on 133–4
*Al Tanzim* organization 75, 154 n.6
amnesty (1991), civil war crimes 9, 146
anthropocentric perspectives on conflict 23
anti-colonial movements 106
'anti-geopolitical eye' 4–5, 139
Appadurai, Arjun 20
Arab cities, urban geopolitics 20
Arab Deterrent Force (ADF) 68, 82–3, 111, 113–14, 122, 149 n.5
Arab identity of Lebanon 118–19
Arab nationalism 60, 62–3, 139
architecture 11, 14, 17
  Adams on 25
  of enmity 73, 76–80, 106, 108–9, 115, 154 n.1
  geopolitical 115–25
armed irregular militias 14, 21, 71–6, 89, 93, 95, 123, 131, 150 n.1
asymmetric conflicts 14, 19, 21–2, 114, 149–50 n.1
atomic bombing of Hiroshima and Nagasaki 16, 18
Ayn al Rummana 111, 113
  battle of hotels 115–25
  bus shooting 79–80
Azmi Bey 17

Ba'ath party 57, 59
Bacik, Gokham 156 n.1
Balkan wars 22, 102
Battle of the Hotels (*Ma'raka al-Fanadiq*) 10–11, 81–2, 103, 115–25

'being political' process 25–6, 103
*Beirut* (Kassir) 1
Beirut, Lebanon 1–3, 6–11, 14, 20, 24–6, 57–60, 64, 68, 70–1, 74, 77, 81, 85, 89, 99–102, 107, 109–10
  British embassy in 64, 89, 93
  built environment of (*see* built environment of Beirut)
  civil war 2, 6, 24, 26, 31, 79, 114, 123, 132, 146
  cultural travel 33–4
  Davie's study on population 31
  'Golden age' 59, 63, 69–70
  modernization project of 17
  neighbourhoods in 127
  partition of 30, 46, 66, 86, 95, 103, 116, 119
  'politics of urbicide' 27, 102, 104
  spaces for peace 144–7
  Tarazi-Fawaz's study of 34
  urban destruction in 121
  urban geography of 1–2
  urban geopolitics of 50–3, 137–8
*Beirut Madinati* 145–6
Berman, Marshal 22
Bhabha, Homi 129
Bin Laden, Osama 177, 179
Bishop, Ryan 15–16
Black Saturday (Lebanon) 81, 110
'Black September' group 74
Bogdanović, Bogdan 23
  view on urbicide 24–6
bombings of Hiroshima and Nagasaki 16, 18
built environment of Beirut 2, 14, 24, 26, 103, 105–8, 114, 123, 136, 142–3
  destruction of 22–3
  in political violence 21–2

and propagation of violence
102–8
semiotic/symbolic interpretations
27
and sovereignty 131–6

Cairo Agreement in 1969 67, 73–8,
82
cartography/cartographers 31, 36,
40, 146
military and commercial 38–9
Catholic and Protestant missionaries
32
Cedar Revolution 127, 144
Chamoun, Camille 57, 59–60, 62,
64–5, 77, 176
Chehab, Fouad 70, 72–3, 76
Christian community in Lebanon 97
Christian Maronite 34, 56
citizen/citizenship 14–15, 25–6,
37–8, 59
city vs. countryside 25
Clancey, Gregory 15–16
coastal road massacre 148 n.2
Cold War 13, 15, 18, 20, 53, 57–8, 60,
72, 102, 141, 177, 179
colonial city, destruction of 16–17
colonialism 28, 30–1, 33
colonial modernity 34, 46, 140
communal homogeneity 51
communism 59, 62–3
Communist revolutionary movement
116
confessional society 150 n.1
consociation democracy, Lebanon
56, 58
The Constitutional Bloc 55
contrapuntal reading, Said's 6, 101,
124
conventional sovereignty 137
Corm, George xii, 26
cosmopolitanism 34–5, 59
vs. sectarian chauvinism 25
Coward, Martin 22–3, 27, 70, 102
critical geopolitics 4

cultural imaginary 52
European 35
cultural travel, Lebanon 33
'culture of sectarianism' 34

The Damascus Agreement 82
Davie, Michael 31, 51, 114–15
decentralization 18, 152 n.6
de facto sovereignty 6, 64, 109, 124,
127, 129, 135, 137
Deir Al Qamar, Lebanon 31, 39, 45
demarcation 16, 18, 39, 103
De Nerval, Gerard 33–4
The Deuxieme Bureau 72–3
*Die Fälschung* (*Circle of Deceit*) film
86
Dijkink, Gertjan 4
dispositif 30, 33, 150–1 n.2
sect as 42–7, 140
Document of National
Reconciliation (Taif
Agreement) 148 n.1
domestic sovereignty 64, 128
Double Kaymakamate of Mount
Lebanon (1842–1860) 37–42,
47
Druze 29, 34–5, 39–40, 43, 45, 48,
141, 153 n.15
Dufferin, Lord 44–6

East Beirut 33–4, 106, 108
economic policies, Chamoun's 60
Eisenhower Doctrine 57, 59, 62, 86,
153–4 n.3
Eisenhower, Dwight 59–60, 65, 176
administration 63–4
El-Khazen, Farid 67, 69, 110
Europe
-backed policies 50
bombing of European cities 18
city-states/free cities 15
colonial cities of 16–17
debate on urbicide 23
modernity in 16
European Commission 48, 50, 52

Farish, Matthew 18
'Fatahland' 74, 90, 155 n.3
*Fighters for Peace* organisation xii, 9
*Film Ameriki Tawil* (film) 148 n.1
First World War 15, 17, 29, 34, 47, 51, 53
Fleming, Ian 176-7
Ford, Gerald R. 10, 91, 95
foreign policy towards Lebanon 4, 7, 9-10, 57-8, 95-9, 101, 109, 139-40, 149 n.5
Forensic Architecture research agency 24
former fighters/combatants xii, 10, 105
  Abu Layla xii-xiii, 107, 117
  Edouard xiii, 75, 79, 108, 114-15
  interview of 10-11
  Nizar xiii-xiv, 79, 106-7
  Omar xiii, 24, 102-3
  Rashid xiv, 106, 110
Foucault, Michel 9, 150 n.2
  power/knowledge mechanism 5-6, 149 n.4
  subjugated knowledges 5-6, 102, 114
'Fourth Sector' (*al qita' al Ruba'*) 81
The French mandate and the Lebanese Republic (1920-1946) 47-9

genealogical geography 39
geopolitical scripts of Lebanon 57-63, 86-7, 101, 113
  chaos script 93-6
  fate script 96-100
  tragedy script 88-93
geopolitics 1-4, 19, 21, 102. *See also* urban geopolitics
  cities and 14
  critical 4
  de-subjugating 138-9
  Dijkink's description of 4
  geopolitical architectures 115-25

geopolitical imaginations 5, 20, 80, 82, 93, 95, 102, 106, 109, 121, 141
  mainstream 101, 114-15, 124, 127, 143
  militia 148 n.2
  official 9, 64, 101-2, 108-9, 124, 137-9, 142
  pacific urban 142-4
  regional 1, 3, 10, 53, 127, 132, 177
  of statecraft 101
  subjugated 6, 102, 114
  urban 3, 14, 19-21
globalization 19, 124, 129
global politics 4, 102, 131, 139
  urban space of (*see* urban space of global politics)
'Golden age,' Beirut's 59, 63, 70
Graham, Stephen 15, 20, 24
Greater Lebanon 29, 34, 45, 48-50, 52
guerrilla warfare 3, 13
Gulf (oil economies) 59, 63

Hanf, Theodor 60, 70, 75
*Harakah al Wataniyya* (National Movement) 75, 81-2
Hariri, Rafiq 144
  assassination of 127
Heidegger, Martin 27, 102
heterogeneity, Coward's on 27, 102
heterogeneous manifestations 5
Hewitt, Kenneth 18
Hezbollah political party 87, 131-5
Hiroshima and Nagasaki, bombing of 16, 18
'historical (sectarian) communities' (*communautés historiques*) 49
Holiday Inn hotel, battle of 115-25
'homeland' 15, 18-19
hostility 24, 32, 59, 109, 113-14, 178
Hourani, Albert 34
hybrid sovereignty 71-6, 127-8, 133, 135-6, 156 n.1, 178
  epistemology 130

hybridity 129–30
lack of sovereignty to 128–30

imperialism 32, 36
'indirect displacement' 22
international law of La Hague (1907) 16
International Relations and Political Science 13
international sovereignty 131
interventionism 62, 87, 99
intra-state urban warfare (7 May clashes) 11, 127
irregular militias. *See* armed irregular militias
Isin, Engin 25–6, 103
isolationism 155 n.9
  isolationist militias 81, 110, 119
  isolationist policy 155 n.9
Israeli Defence Force (IDF)/Israeli army 114, 148 nn.2–3, 149 n.3
  in Lebanon 3
  against Palestinian 24
Israel/Palestine peace process 87, 99, 128

*Jabal Lubnan* (Mount Lebanon) 29
Jackson, Robert 128–9
Japan, bombing of cities in 18
Jibril, Ahmad 121
Joseph, Suad 69–70

Kassir, Samir 1, 20, 60, 62–3
Kata'ib (*Phalanges*) militia 75, 116
Kata'ib party 155 n.9
Khatib, Ahmad 74, 82
*Khatt at-tamas* (Green Line) 116, 119
Khoury, Bishara al 55, 57
Kissinger, Henry 87, 91, 96, 98
Koleilat, Ibrahim 109–10, 114, 119, 121

Lacoste, Yves 101
La Hague (1907), international law of 16

laic education 32
*Laïque Pride* movement 144–5
late-Ottoman Lebanon 10, 30–2, 34
League of Nations 152 n.14
Lebanese Armed Forces (LAF) 93
Lebanese Front militias 81, 106, 110–11, 113, 116
Lebanese nationalism 109, 114, 119, 124
Lebanese National Movement militias 74
Lebanese Phalanges 81–2, 156 n.4
Lebanese Security Forces 100
Lebanon 17, 25
  Arab identity of 118–19
  Beirut (*see* Beirut, Lebanon)
  civil war xii–xiii, 1, 4, 6, 66, 89, 91, 93–4, 150 n.3
  Deir Al Qamar 31, 39, 45
  double imaginary of 34–5, 42
  foreign policy towards 4, 7, 9–10, 57–8, 95–9, 101, 109, 139–40, 149 n.5
  former fighters (*see* former fighters/combatants)
  geopolitical scripts of (*see* geopolitical scripts of Lebanon)
  IDF/Israeli army into 3, 148 n.2
  late-Ottoman 10, 30–2, 34
  modernity/modernization of 29–33
  non-intervention in 85–93, 95, 97, 99, 101
  partition of 30, 37, 40–2, 52
  political violence of 25, 31
  Rasheya 45
  refuge and battleground 33–7
  reorganization of territory 46
  socio-economic perspectives on violence 68–71
  sovereignty of (*see* sovereignty of Lebanon)
  Two Years' War (*see* Two Years' War (*al-harb as-sanatayn*))

Maasri, Zeina 111
'*Macrocéphalie Beyrouthine*' 70
mainstream geopolitics 101, 114–15, 124, 127, 143
Makdisi, Ussama 32, 34, 39, 43, 150 n.2
*MARCH Lebanon* organisation 9
Maronite 29, 34–5, 40, 43, 48, 52, 55–6, 141
May 2008 clashes 11, 127, 144
  sovereignty and built environment in 131–6
Mediterranean Levant 29, 33, 49, 59, 63
militarization, urban 21, 143
militia combatants/fighters. *See* former fighters/combatants
militia geopolitics 148 n.2
militia warfare 101, 103, 107
Mitchell, Timothy 33
modernity/modernization, urban 15, 22
  in Europe 15–16
  of Lebanon 29–33
  project of Beirut 17, 24
'Mostar '92 – urbicide' 23
Mount Lebanon 29–30, 33–6, 38–40, 42, 45–6, 50–1, 53, 56, 176
multi-communal society 150 n.1
Multi-National Force in Lebanon (MNF) 155 n.2
Murphy, Robert 63–4
The Mutasarrifiyya of Mount Lebanon (1860–1920) 42–7

Nasrallah, Sayyid Hassan 131–2
The National Bloc (Christian political group) 55
nationalization of Suez Canal 59–60
National Movement militias 110, 116–20, 121, 123
The National Pact (*al mithāq al watani*) 55–8
nation state 3, 7, 13, 15, 18–20, 29–33, 38, 46, 48, 56, 92–3, 96, 99, 102, 110–11, 123–4, 139
natural inclinations 98

negative sovereignty, Jackson's 128
neutral nation 72, 76
new municipal building laws (1986) 17
'new wars' 13, 19–22
non-intervention in Lebanon 85–7, 95, 97, 99, 101
  as territorial trap 88–93
non-sectarian social geographies 39
non-state armed actors 2, 10, 15, 21, 89, 99, 101–2, 110, 114, 130–1, 135–6
non-state political violence 20–1

official geopolitics 9, 64, 101–2, 108–9, 124, 137–9, 142
O'Kane, Maggie 5
Operation Blue Bat 59, 62, 94–5, 127
  urban geopolitics of 63–6
Operation Litani, Israeli army 3, 148 nn.2–3
*Orientalism* (Said) 86
Ottoman-Egyptian war (1831–1840) 29, 37–8
Ottoman Empire 17, 25, 30–1, 33–8, 46–9, 51, 53, 151 n.6
Ó Tuathail, Gearoid 5, 89

pacific urban geopolitics 142–4
Palestine, IDF against 24
Palestine Liberation Organization (PLO) 72, 74–8, 82, 148 n.2, 178
Palestinians
  armed groups 74–5
  guerrillas 74–5
  into Lebanon 72
  population 73
  resistance 73, 75–6, 78, 148 n.3
Pan-Arabism 56–7, 59, 72, 86, 109, 113–14, 119, 124
paramilitary violence 103
Peres, Shimon 90–1, 93, 97–8
political geographers 88, 124, 128, 138
political geography. *See* geopolitics

political poster 111, 113–14
    by Al Murabitun militia 119–22
    by Arab Socialist Union 61
    Lebanese Forces propaganda 112
political sectarianism of Lebanon.
    *See* sectarianism
political violence 14–15, 20, 25,
        30–1, 52, 110, 118, 131, 135–6,
        141–2, 146
    Bogdanović on 24
    Coward on 23–4
    non-state 20–1
    against urban built fabric (*see*
        urbicide (city-killing))
'politics of urbicide,' Coward's 27,
        102, 104
Popular Committees 123–4
post-Cold War 19–20
    urbicide 22, 27
post-war reconstruction 3
power/knowledge mechanism,
        Foucault's 5–6, 149 n.4
pre-*Tanzimāt* Mount Lebanon 39, 43
property and material damage 25
*Protein* (Sidon-based company) 77

Rasheya, Lebanon 45
'rational superiority,' European 42
refuge/refugees 56
    and battleground 33–7
    camps 11, 71–3, 78–9, 81, 98, 142
    Palestinian 51, 72–3, 81
    settlement patterns of 51
    Shi'i 60
regional geopolitics 1, 3, 10, 53, 127,
        132, 177
Reglement (*düzeni*) of Şekib Effendi
        41–2, 44
*Réglement Organique* administrative
        document 47
religious community (Taifa) 30, 33,
        36–7, 47, 49, 52, 56, 58, 150
        n.1, 153 n.15
religious/sectarian violence 32
'repositioning,' political 19

Republic of Greater Lebanon (*Grand Liban*) 48
rural areas 24–5, 71–2

Saad, Marouf 77–8
Said, Edward 6, 86
San Remo Conference (1920) 47
Sassen, Saskia 19
Sauvagnargues, Jean 90, 92
Second World War 16, 18–19, 55, 58
sect
    as dispositif 42–7
    as modernity 29–33
    normalizing 47–9
    as territory 37–42
    unavoidable 96–100
sectarianism 30–3, 46, 56, 70
    decolonizing 140–1
    identity 156 n.5
secular education 32
Shehadi, Nadim 132
Sidaway, James 128–9
The Sidon fishing protests 77–8
social revolt 43
socio-economic perspectives on
        violence 68–71
socio-environmental disasters 16
Solh, Rashid el 155 n.9
Sorkin, Michael 20
sovereignty of Lebanon 57–8, 62–3,
        65–7, 77, 86, 88, 91, 138–9
    approaches of 128–9
    and built environment 131–6
    de facto 6, 64, 109, 124, 127, 129,
        135, 137
    domestic 64, 128
    hybrid (*see* hybrid sovereignty)
    international 131
    reframing 108–15
sovereignty regimes, Agnew's 129
sovereigntyscapes, Sidaway's 128–9
Soviet Union 66, 87
state and irregular armed groups 71–6
statecraft, geopolitics of 101
state sovereignty model 89

Stewart, Desmond 64
subjugated knowledges, Foucault's 5–6, 102, 114
suburbia, planning 18
Suez Canal, nationalization of 59–60
Sunni Muslims 37, 48, 52, 55–6, 78
Sykes-Picot agreement 48, 152 n.13

*Tanzimāt* (reorganization) of Ottoman Empire 37–9, 46, 51
Tarazi-Fawaz, Leila 31–2, 34, 51, 151 n.4, 152 n.6
techno-centrism, urban geopolitics 21
territorial nation state 92
territorial trap 95, 139
  Agnew's term of (assumption) 108–9
  of geopolitical scripts 115
  non-intervention as 88–93
Thompson, C. W. 151 n.3
Truman Doctrine 59
Two Years' War (*al-harb as-sanatayn*) 1–3, 25, 30, 68, 80–3, 94, 100–1, 109, 116–17, 123–4, 127, 136, 138, 140, 142, 146, 148 n.3
  geopolitical knowledges of 6
  rounds (*Jawlat*) of fighting 4
  urban geopolitics of 21

Umam organisation 9
United Arab Republic (UAR) of Syria and Egypt 56–7, 59–60
The United Nations 149 n.3
United Nations Interposition Force in Lebanon (UNIFIL) 149 n.3
United Nations Relief and Works Agency (UNRWA) 72
The United States 22, 58–60, 62–3, 66, 87, 90, 95, 101, 123, 153–4 n.3, 154 n.4, 179
  US Marines intervention in Lebanon 65
  US non-interventionism 87
  US Sixth Fleet 60, 62, 86

urban architectures of enmity 76–80. *See also* architecture
  bus shooting and start of war 79–80
  the Sidon fishing protests 77–8
urban geopolitics 3, 14, 19–21, 31, 48, 113, 127, 139, 176–7, 179. *See also* geopolitics
  of Beirut 50–3, 137–8
  of Operation Blue Bat 63–6
urban guerrilla warfare *vs.* airpower 3
urban space of global politics 14
  conflict and violence 14–16, 21
  decentralization 18
  modernity 15
  town planning 18
  *vs.* rural 24–5
  Western features' 17
urban territory 8, 67, 99–100, 104, 114–15, 127, 137
urbicide (city-killing) 19, 22–8, 121, 124
  of Beirut 138
  city of 24–8
  contextualizing 141–2
  Coward's works on 70, 102, 104
  genealogy 22–4
*Urbicide: The Killing of Cities?* (symposium) 23

Vilayet of Beirut 48, 51
violence 25, 27, 48, 62, 80–2, 98, 148 n.1, 150 n.3, 151 n.4
  attack on Al Mustaqbal newspaper building 133–4
  attack on religious buildings 150 n.3
  built environment and 102–8
  paramilitary 103
  political (*see* political violence)
  religious/sectarian 32
  socio-economic perspectives on 68–71
  Tarazi-Fawaz on 34
  urban 21
Von Metternich, Klemens 40

The War on Terror 13, 20
Weizman, Eyal 24
West Beirut 33–4, 106, 121, 131–2
Westphalian (or state sovereignty) model 88, 124, 138
Wogensky, Andre 115–16

Younis, Kamal 118–19
#*YouStink* protest movement 144–5
Yugoslavia, former 23, 27
Yugoslav National Army 23